THE FARM

Farming in the South West

Whereas generally the differences in Britain between farming in the late-Victorian period and the Edwardian period are minimal, the differences in agricultural practices between the south west and the rest of England are marked. This is nowhere better illustrated than in the pages of William Marshall's *The Rural Economy of the West of England* (1796). Marshall terms the farming of the south west 'Damnonian Husbandry' after the ancient name for the kingdom of Dumnonia (once encompassing all of Devon, Somerset and west Dorset) and he draws attention to its 'peculiar' elements. First he remarks on the prevalence of small farms – something that has survived well into the 20th century. He goes on to discuss the various breeds of cattle, sheep and pigs, and how they are kept and are 'peculiar' to the region, while also commenting on how the management of crops from sowing to harvesting and threshing, along with the tools employed are, again, 'peculiar'. He concludes with the line: 'But what strikes us most forcibly… is, that, in the lapse of centuries, its Rural Practices should not have assimilated, more freely, with those of the Island at large.'

These differences in agricultural practices carried on into the Edwardian period and can, in places, still be seen today. The major reason for the continuity and distinctiveness of Damnonian Husbandry is cultural, as successive generations have passed down tried-and-tested methods. Take, for example, the West Country shovel. Shaped more like the 'spade' in a pack of playing cards, it is very popular in Devon. Peter and I struggled to get to grips with the longer handles and the tipped blades, but the locals were quick to assure us of its superiority over the short-handled, flat-headed shovel we were more familiar with.

The geological formation and climatic conditions of the south west have also governed the way in which the landscape of Devon and Cornwall has been farmed. The peninsula of south-west England is dominated by a backbone of granite rock, which in Devon forms Dartmoor and in Cornwall, Bodmin Moor. These harsh and unforgiving moors, along with Exmoor to the north, have played a crucial role in animal husbandry, while the red sandstone of the Old Devonian, shale and thin sandstones of the Culm Measures support rich fertile soils.

Perhaps most influential, as far as farmers are concerned, is the climate. The North Atlantic Drift and the Gulf Stream bring warm winds and currents from the Gulf of Mexico. Cornwall and Devon benefit from what can feel, at times, like tropical conditions. Even in winter, the temperature can be relatively mild and spring comes much earlier. Famously though, the Gulf Stream brings rain with it; Cornwall and Devon record some of the highest annual rainfall figures in Britain. But rain can be a good thing and the early springs and the summer rains have been harnessed by West Country farmers to grow early fruits and vegetables, while traditional dairying practices have benefited from the lush pasture created by the warmth and rain.

The patchwork quilt effect of the field systems of the south west of England.

This unique combination of geology and climate, coupled with the long-term continuity of traditional farming practices – in both William Marshall's day and during the Edwardian period – set Cornish and Devonian farming apart from that of the rest of England.

THE TRADITIONAL DEVONIAN FARM

One of the most striking characteristics of the rural landscape of the south west of England is the frequency of small farms and it is clear from historical and archaeological sources that the Devonian pattern of small enclosed farmsteads dates back to the 14th century and beyond. In medieval times the open field system prevalent throughout the Midlands, East Anglia and central and southern England consisted of a small number of large open fields within which all the village occupants would work a number of strips of 'ridge and furow' and would, through collective cooperation, maximize their productivity. It was widely believed that in the extremities of the British Isles, such as the south west of England, this advanced method of farming simply hadn't evolved. It is now thought that Devon and Cornwall had, in fact, developed its own advanced form of husbandry and one far better suited to the regional terrain and climate. This type of husbandry relied less on the 'nucleated' village where houses are clustered in one place with the inhabitants working open fields 'in common' and more on small dispersed farmsteads in family occupation, each with their own enclosed fields systems.

A characteristic of these small farms is the Devon longhouse. A typical example of this would be rectangular in plan form. It would have an entrance at a point roughly halfway along one side with a through passage to an opposing external door. Internally, the building would be divided between livestock on one side and human occupants on the other. The 'hall' would be the main living area for the farmer and his family with a hearth set either against the end wall or within an internal dividing wall. 'Shippon' is the Devonian word for a cattle byre and with an open gulley running down the centre, it was always situated down hill of the hall so that animal effluent could flow away from the living space. Above these rooms in the roof of the longhouse would be, respectively, the sleeping quarters and the hay loft.

This arrangement of rooms dates back to the medieval period and it is one that can be found along the Atlantic seaboard from the Highlands and Islands down through Wales and Ireland, the south west of England

PRINTED & PUBLISHED BY THE UNITED COMMITTEE FOR THE TAXATION OF LAND VALUES. 20. TOTHILL ST LONDON S.W.

THE ICONIC PATCHWORK QUILT OF THE DEVON LANDSCAPE

The fields of the dispersed yet self-contained farmsteads of the Devon landscape are small and enclosed by hedge banks. These are banks of earth faced in stone and with a hedge planted on top. The result of this arrangement of small enclosed fields grouped around a courtyarded farm creates the appearance, when seen from a distance, of a patchwork quilt of greens, browns, yellows and reds (depending on the season). Rider Haggard in his *Rural England* (1902) commented, 'perhaps no English county that I have seen is quite as lovely as this land of Devon' and the modest size of the enclosed fields gave the impression that 'here the land is valuable – a desired possession'. It is true that on a summer's day the verdant and lush landscape of Devon with its iconic chequerboard landscape can enrapture the heart.

and into Brittany – in fact any place where small regularly spaced farmsteads prevail over nucleated villages and outlying fields. In time various additions would be made to the longhouse with lean-tos against the hall and rear of the building adding additional space for living quarters, 'wet' rooms and cold stores. Further buildings such as stable blocks, threshing barns and cow sheds might, certainly by the 19th century, be built around a cobbled courtyarded area to create an enclosed farmstead with a pond and stack yard completing the idyllic scene.

Idle land means idle men: a poster produced in 1910 for the London County Council by the United Committee for the Taxation of Land Values.

DARTMOOR

The Tamar is a steep-sided valley and down in our cottage it is easy to forget the outside world. But if you walk up the hillside – and it's quite a climb – there from the top you look out over Dartmoor, only a couple of miles distant. It's hard to overstate the contrast. In our valley the soil is deep, rich and black; the air is still and damp in the winter, humid as the sun warms it. The south-facing slopes ripen some of the earliest outdoor strawberries in Britain. Wherever the land is left uncultivated trees grow fast and tall. The river is a highway out to the markets of Plymouth and, via the railway, London, Manchester, Edinburgh and beyond.

A handful of miles away the people of Dartmoor live a radically different life. The soil is thin and acid; the wind blows strong all year. Trees struggle to gain a foothold, growing slowly where they grow at all. High, bleak and clear, you can see for miles. Holdings up here have to be large and the people are thinly scattered.

Despite being so close to one another, the two regions farm in completely different ways. Market gardening dominates our prospects, with potatoes and grain grown to feed us and our animals; up on Dartmoor the mainstay is raising cattle. Sheep, ponies and rabbits were the other main products of the moor with a small wild whortleberry crop in the early summer.

CATTLE ON THE MOOR

Beef prices were good in the Edwardian period, holding up well when other agricultural prices were falling. The cattle raised on the moor could be produced quite economically if they were allowed to range over large areas of common land – but such grazing could

not produce the final product demanded at the butcher's shop. Once they were two or three years old the young cattle were rounded up and sold on to graziers with richer pastures at lower altitudes where they were fattened up ready for slaughter.

During the harsh Dartmoor winters the cattle did best if they were brought indoors and fed on hay, with readily available sedge cut from the moor used for bedding rather than scarce straw. Down in the valleys cattle were often fed turnips and mangel roots through the winter. Dartmoor cattle, however, were more likely to be fed on oats and oat straw to supplement the hay – turnips not doing so well on the moor in the thin impoverished soil.

When March came around many Dartmoor farmers set fires on the moor to burn off patches of old heather and encourage the growth of new rich grass for their animals. This practice was known as 'swaling'. The cattle were turned out to roam the moor as soon as the weather permitted. When out on the moor the herds could largely fend for themselves. A rough eye was kept on them by the 'moormen' who steered them clear of the artillery ranges and rounded up those that strayed too far from their traditional grazing areas, each herd having its own territory.

Each herd of cattle on the moor largely keeps to its own territory, without the need for man-made boundaries.

The Tamar Valley, with its steep sides and limited view-sheds, can often get claustrophobic. When I climbed up to the hills to take in the wider view, Dartmoor would always loom large on the horizon to the east. Early in the year, I was itching to get up there and explore its wilderness. I knew very well that for centuries the moor had proved an invaluable source of grazing for the farmers of the parishes that border its remote fringes. I was also aware of the many other industries that this barren wasteland supported, such as mining, quarrying, peat digging and the pony drifts. Yet, it was the place Dartmoor occupied in the literary history of the south west that most beguiled me. Sir Arthur Conan Doyle famously set his most popular detective story on Dartmoor in which his chief protagonist, Sherlock Holmes, grapples with the mysteries of the 'Hound of the Baskervilles'. I believe what makes the tale so popular is that so much of it was steeped in the myths and legends of the moor, drawing together a whole host of ancient tales into one thrilling yarn.

RABBIT FARMING

Rabbits were another important form of livestock on the moor. There was a healthy market for rabbit meat in the hotels and restaurants of Devon, thanks to the wealthy tourists who visited the county. Dartmoor rabbits were also whisked by rail to the poultry merchants in Birmingham and Sheffield. Rabbit fur was also in demand.

Rabbit warrens were large commercial concerns, several of them covering a little over a thousand acres. William Crossing in his 1903 articles in the *Western Morning News* lists ten warrens on Dartmoor, although not all of them of that size. Artificial burrows of stone and earth were built to house the rabbits; vermin traps were dotted about to protect them from predators, and small hides allowed men to keep watch at night for foxes and poachers.

To make the burrows, the warreners first dug a narrow trench about 16 feet (4.8m) long and about a foot (30cm) deep, with a series of small branches off it, not opposite one another but staggered. Large flat stone slabs were laid across the trenches to form roofs on the runs, then rubble and soil were piled on top to insulate the rabbits from the cold. In a harsh winter the rabbits were fed on furze and hay cut from the moor, but mostly they were left to graze.

From March to August the rabbits were allowed to breed and fatten. Through the rest of the year the warreners trapped the rabbits, moving from one area of the warren to the next in turn. Some used snares but many preferred nets and ferrets. Nets were set at all the entrances to the burrow and a ferret sent in. Fleeing from the ferret, the rabbits ran out and into the nets where they could be quickly dispatched. Other warreners preferred to use nets and dogs, setting long walls of nets between the feeding areas and the burrows while the rabbits were grazing. A commotion of dogs and men then sent the rabbits flying back to their burrows only to be caught in the lines of nets. Either way, nets usually meant a much quicker end for the rabbits than snares.

At the end of the 19th century rabbit fur was no longer so fashionable and the warrener was now farming for the meat alone. This drove a couple of the warrens out of business; there was pressure too from small farmers all over Devon taking up rabbit farming – advice being available in the Board of Agriculture and Fisheries leaflets.

PEAT

Wood is in short supply up on Dartmoor and carrying coal up from the river, canals or railway depots to the scattered dwellings of the moor was a very expensive business. While cheap coal provided the common domestic fuel in our valley, the homes up on the moor were fuelled with peat. In the past peat had been commercially extracted for industry, but by the Edwardian period it had reverted to local domestic use.

Each farmer had his own 'tie' or pit where he cut the family's peat, high on the moor where the deposits are thickest. He began with a single narrow trench, 40 yards (36m) long and two or three feet (60–90cm) deep; then he sliced the peat from the side of his trench, widening it into a pit over the years. Using a long knife, a budding iron and a turf iron, the peat was neatly cut into blocks 20 inches deep, 14 wide and 2 inches thick (50 x 35 x 5cm). These blocks or slabs were dried before they were carried away. Firstly they were leant against each other in pairs like a pair of playing cards; later as they dried they were formed into small piles called 'stooks', rather as wheat and barley were on arable land. Once dry the peat was carried to the farmhouse and either stacked under-cover or thatched like a rick. If it were to become sodden it would be useless.

Using peat as a domestic fuel had an impact on the home life of Dartmoor folk. To begin with there is the smell. Peat is an especially pungent fuel – most people find it very pleasantly so. Someone from a peat-fuelled Dartmoor home would have smelt quite different from someone from the sulphurous coal-fired valleys. Peat burns slowly and gently as an open fire, but is not suitable for using in enclosed iron ranges. Cooking on peat is a matter of long slow simmering or gentle grilling, which of course suits some recipes better than others. The smoke imparts its flavour to any food that it comes into contact with. Dartmoor clotted cream was especially popular with many visitors for this reason (see also page 148).

Without an iron range, Dartmoor homes continued to use the old fashioned 'clom' ovens for baking. These are fired clay domes built into the side of the fireplace. They are heated by lighting a fire inside of small dry sticks. The fire heats the clom oven and the stones around it. The ashes are raked out and the hot oven will then bake whatever is put into it. Dartmoor luckily supplied a ready source of free fuel for clom ovens in the form of old woody heather and furze bushes.

Alex bringing home an early morning catch of rabbits.

WHORTLEBERRIES

At Tavistock market in 1903 whortleberries were fetching 6d or 7d per quart. The wild berries, often known locally as 'hurts', grew in abundance on Dartmoor and gathering them for sale almost counted as a holiday for many people who lived on the fringes of the moor. A day's pay could be earned as you enjoyed the sunshine and views of a summer day up by the tors.

DRY-STONE WALLING ON DARTMOOR

While hedgerows and hedgebanks proliferate in the valleys, an altogether different form of boundary is used on the foothills, commons and moors of the Devon landscape. Mankind has employed stone in the construction of settlements and ritual monuments since the Bronze Age and earlier, and it is certain that before Roman Britain, stone was used to build stockproof walls. Early walls are extremely difficult to discern in the landscape without detailed archaeological research, not least because the stone has been reused in successive periods of wall building.

In accordance with the medieval laws of the 'Forest' of Dartmoor, settlers could enclose land in an attempt to domesticate it and improve it for a range of agricultural purposes. Dartmoor farmers have taken up these rights of enclosure in varying degrees using dry-stone walls to create a hardy stockproof boundary for the newly enclosed land. Hedges struggled to grow on the windswept hillsides of the moor; timber to build fences is in very short supply – but stone is in abundance.

RIGHT (from left to right) The Devonshire shovel, sledge hammer, three 'bar ire', a West Country shovel and a mattock or 'vizgie' – all tools used by the waller to move the larger stones into place.

FAR RIGHT Jason leavers a large stone with the 'bar ire' – as it was known locally whilst I insert a levelling stone.

Any visitor to the moor today can see field systems of varying character enclosed by stone walls. Some appear as small islands of fields surrounded entirely by the open moor. Others seem to be clearly an extension of a field system lower down the moorland slopes where farmers have attempted, in a piecemeal fashion, to take under cultivation land that was once wild open moor. The overwhelming majority of these walls come from that period of optimistic land 'improvement' in the late-18th and early 19th century. In many cases this optimism was misplaced and these landscapes of acidic, stony and coarse ground never returned the capital funds invested in them. On Dartmoor, such attempted improvements are known as 'landyokes', 'landbotes' or, more commonly, 'newtakes' and the walls of these enclosures stretch ambitiously, in a ruinous state, over some of the remotest parts of the moor.

This was a time in Britain's agricultural history when it was felt firmly that improvements could be made to even the most barren stretches of land – from the mountains of Scotland to the moors of the south west.

SHAPES AND STYLES OF WALLS

Walls built high up on the moors and made to control the movement of sheep might be quite thin and care would be taken to ensure that relatively large gaps existed between the stones so that the strong winds blew through them rather than pushing the whole affair over. Some walls would have a vertical dressed face on one side and a sloping earth bank on the other. Such walls represented the boundaries between the open moor and the surrounding farm land and were designed to ensure that stock – such as ponies and sheep – moved in only one direction and that should they manage to scale the vertical dressed face of the wall, they could easily be driven back on to the moor.

OPPOSITE Working on Dartmoor's walls was fun and fast. The end product didn't have to be too pretty, just functional, so we managed to quickly get a considerable stretch of wall repaired.

There are, however, many walls that served a purpose on the moor in the Edwardian period and I was fascinated to learn of the many types of dry-stone wall that can be found still functioning in the Dartmoor landscape of today. These various styles are determined by the types of stone found across the moor. Although Dartmoor is essentially a huge mass of granite, variations in the natural geology mean that the stones quarried from different areas are not only different in colour but also in the way they behave when 'dressed' to form part of a stone wall. Differences in style of building also come about with different periods of history. In very early walls, the builders worked much more with the shapes of the stones but by the 19th century builders had the tools to dress the stones to more even shapes – resulting in stronger and more permanent structures. Finally, the function of the wall would ultimately determine its style of build.

In my experience, stone wall building is one of the best ways to spend time in the great outdoors. Although slow and laborious, the end results are satisfying. So, having learnt a little about the history and styles of Dartmoor's stone walls, I was keen to learn about the practical skills involved in what must have been one of the moor's first labours as mankind attempted to tame the wilderness around him. Wilf Hutchins, Jason Thomas and Bill Budd proved admirable tutors in this task and they had the lively banter of folk who both enjoy and take pride in their craft. In the first instance I was intrigued to find out that Dartmoor's walls in fact need quite regular maintenance. Harsh winters can cause stones to shatter under the pressure of frost fractures and the frightening gales that tear across the moor can simply blow sections of walls down. Animals pass into the gaps and as cattle and sheep scratch and rub at the walls they, in turn, help to bring down stones. The roots of trees growing close to the walls can cause upheaval, and finally, careless ramblers scaling walls can dislodge capping stones. These might seem like long-term processes, but on such a vast scale it means that there is always work for the waller to do. It wasn't long before Wilf, Jason and Bill had me stuck into some crucial repair work and I found myself instantly immersed in the 3D puzzle that is dry-stone walling. It takes a while, though, to engage with the stones, their shape and distribution along the proposed face. The weather that day was particularly favourable and I have to admit, if it wasn't for the pressures of the farm awaiting me in the valley bottom, I could have stayed up there all month.

IN THE FIELDS

POTATOES

Potatoes became a regular feature of crop rotation in the early 20th century. They helped enormously to increase the output per acre and per man as a cash crop. 'Early' potatoes fetched the most money, and growing what we today call 'new' or 'salad' potatoes really took off in the south west. Bringing the harvest forward by a week could increase a farmer's income by as much as £10 per acre: instead of getting between £25–£35 per acre a grower could get up to £45 – a real bonus.

Chitting and Planting Potatoes

I decided I was going to grow a potato crop, as even farmers who didn't specialize in the market would have had a patch for their own use or for local sale. I discovered that our local hay merchant, Mr Jim Hamilton, was also an expert potato grower. He told me to start by 'chitting' my seed potatoes, a technique where the potatoes are exposed to warmth and light to promote early sprouting. Chitting gives the potatoes a head start before they are planted. However, it isn't without its problems. The sprouts easily break off and so the potatoes couldn't be planted by machine. Also, if they were planted with sprouts facing down into the ground, time would be wasted as they tried to grow upwards or – worst still – died. No, there was only one way in which the potatoes could be safely planted, and that was by hand. I had set aside roughly two thirds of an acre for potatoes, and a

Despite the many technological, scientific and mechanical breakthroughs of the hundred years running up to the Edwardian period, potatoes in the early 20th century were still being sown by hand.

great deal of work needed to be done. The ground needed thoroughly working down to a significant depth. This involved endless cultivating and harrowing crossways over the field, so that the ridging plough could be drawn freely through the soil, creating straight drills. The ridging plough has a mould board on both sides so that it cuts a deep channel and 'banks up' on both sides, leaving a series of ridges and troughs between two and three feet (60–90cm) across. We shovelled manure into the troughs (or drills) – no other crop needs as much fertilizer as potatoes. We spread a full 20 cart loads of our farmyard manure across the patch and, after an exhausting day, had a leisurely morning of setting potatoes into the manure.

Then came the really hard bit. Ridging up alone had proven extremely difficult and getting a straight line was proving beyond me. It seemed that at a certain point in the field the plough seemed to float to the left and then to the right before straightening up again and this had the effect of producing a kink in every drill. The archaeologist in me was sure that this was due to the plough hitting some ancient bank or ditch running across the field but I could see that Peter and Ruth weren't so convinced. In any event, once the manure and potatoes had been laid in the drills, I had to run through the centre of the ridges with the plough and effectively split them so that I cast the earth over the potatoes. I can honestly say that this was one of the hardest things I have ever done in my life. If the plough dropped into the trough instead, I would gouge out the chitted seed potatoes and the manure. I did this a couple of times as I nervously set out but I soon developed a technique that saw me getting the better of the job. Supreme concentration was required along with the strength both to guide the plough and to keep pace with it. If the plants grew in straight lines, I'd know I'd done a good job.

PETER'S DIARY

Ploughing a field throws up a number of challenges, such as getting the horses and the plough to go in a straight line, making sure your furrows match up seamlessly and timing the work with the ever-changing weather conditions, so that you are cutting through ground that isn't too hard or too wet.

Our field also presented us with another challenge – stones. When I say stones I really mean boulders. Basically a ditch had been dug across the area we were ploughing and then back filled so that large stones lay just beneath the surface. Alex first became aware of it when these stones kicked the plough violently into the air – so much so that they snapped off the tip of the iron plough share.

So, while Alex and Megan were ploughing I was probing the area of the ditch (on either side of the field) with a four-pronged muck fork and each time I came across a stone I would dig it out with a mattock and spade.

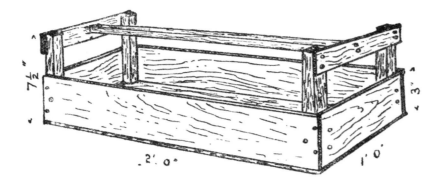

A chitting tray as it appeared in the *Board of Agriculture and Fisheries leaflets. Fortunately for us, local farmer Jim Hamilton still had some stored away in an old barn.*

Potato blight had been a real problem for growers throughout the 19th century and was a contributing factor to the terrible famines in Ireland. Once this fungus-like microorganism takes hold in the upper parts of the plant, it is extremely difficult to stop it spreading into the tubers below. Warm wet weather helps spread the disease: rain washes the spores down into the soil. The disease plagued potato growers some fifty years after the famines and the vegetable's popularity among Edwardian farmers meant that there was a greater likelihood of disease spreading – not just from farm to farm but also through trading seed potatoes.

Knowing the famously wet weather of the south west, I was vigilant in my inspection for the symptoms of blight. These include a burnt appearance to the tips of the leaves which can, if left unchecked, rot the tubers in the ground. The breakthrough treatment against blight and other similar rotting diseases was made in the vineyards of the Atlantic coast of France. One vineyard owner had grown tired of passers-by reaching over the hedge and picking his grapes and he sprayed the fruit closest to the roads with a mixture of lime-water and copper sulphate, in the hope that the bitter taste would dissuade the 'scrumpers'. It worked, and it wasn't long before he noticed that the sprayed grapes were holding out better than his other vines against blight. Thus, Bordeaux mixture was born – and I kept at the ready a jar of copper sulphate and a bucket of lime-powder in case I saw any symptoms.

But a more urgent problem was arising in the field – weeds. Currently the potatoes were out-stripping the weeds in size. It wouldn't be long, however, before the weeds were catching them up and importantly, sucking moisture from the soil that would otherwise be swelling the tubers. I was dreading having to get out into the fields with my hand hoe but Mr Hamilton, my potato-growing guru, came to my rescue with a horse-drawn implement. A 'grubber' or 'scuffler' is a series of hoes fixed to a beam and pulled by a horse. You run it up the centre of the furrows, between the potato plants, and it carves out the soil at the edges of the ridges, dislodging or decapitating all weeds. I was nervous about using it, as I knew that running out of position a matter of six inches (15cm) to either the left or the right would hoe out the very plants I was intent on protecting.

Prince was the perfect horse for the job as his languid pace gave me time to keep up with the implement and direct its course. It wasn't long before we were racing up the furrows and I was

POTATO BOOM

The sheer popularity of potatoes spurred breeders on to find new varieties resistant to blight. This led to the Potato Boom of the early 1900s. Archibald Findley of Auchtermuchty in Fife developed the famously prolific 'Up-to-date' and 'British Queen' varieties in the 1890s, but it was his 'Northern Star' and 'Eldorado' that played a large part in the boom. So successful were these strains in resisting disease that by 1904 there was a potato glut and prices fell. By now there were a perplexing number of varieties and it transpired that certain seedsmen hadn't been completely truthful about their stocks' disease resistance. This lead to the Board of Agriculture setting up a disease-testing centre at Ormskirk, Lancashire in 1914. Government intervention reflects not only the growing importance of potatoes to British agriculture but also their increasing role in the national diet.

LEFT *The Albion potato digger*
was an innovation in potato
lifting. It not only lifted the
tubers from the ground, but
spun them out of the earth
so that they could be easily
picked up.
BELOW *Ruth prepares the*
Bordeaux mixture.
OVERLEAF *Long afternoons*
were spent out hoeing the
potato crop in a bid to keep
the weeds at a minimum.

chuckling to myself as we tore through docks, hogweed, thistles and other nasty field weeds. Although most would wither in the dry conditions, some might very well take hold again and grow on. The important thing was though that I'd given our potato plants a critical advantage and one that would put us in good stead for a harvest in early July. As a finishing touch I ran the ridging plough through the furrows to bank up the earth loosened by the hoeing.

I can't tell you how much I was looking forward to high summer, running the 'lifting' plough through the crop and raising the tubers from the ground. I felt sure that this was, out of all the endeavours of the year, likely to be the most profitable and truly Edwardian. But only when I sat down to a plate of freshly boiled new potatoes dressed in mint and butter would I truly be able to bask in the glory of my potato crop.

THE HUMBLE OAT

Barley, as we all know, can be malted and used to make beer. Somewhat foolishly, I had let my passion for real ale influence my choice of cereal crop for our year on the farm. Both Peter and I harboured ambitions of making a late summer tipple from our own barley but mercifully, with a wry smile and a knowing look, pretty much every cereal farmer in the region steered us away from this extremely difficult to cultivate crop.

Oats, we were advised, were the easiest crop to grow. Both Peter and I can claim direct Scottish ancestry and we both recollect being fed porridge on cold winter mornings when we were kids. Beer or porridge? I know which I'd rather, but the more I read about oats, the more this most humble of crops began to appeal to me.

Historically, there were good reasons for choosing to grow oats at Morwellham. The cereal, has the ability to support the strength of horses – particularly heavy horses – and in the Edwardian age there were more working horses in Britain than in any other period. There was therefore a great demand for oats. Whereas the acreage of wheat grown plummeted between 1885 and 1895, the number of acres being sown with oats increased. Another attractive feature of the plant is its versatility. From the strongest clays to the lightest sandy soils, oats can be sown throughout the country. And, with a vigorous growth habit, they will tend to choke out most weeds.

Harvesting the Oats

The oat crop was the last harvest of the year. And a lot depended on it. All in all, the project had gone really well – so far. Despite the wretchedly cold winter and the frozen ground, we had managed to get the ploughing done and a prolonged dry period in early March had allowed us to work down the ground into a fine tilth. By this time the horses were well practised in their art and sowing the seed went without incident. A few weeks later, after some light showers, the plants sprouted up and were beautifully aligned to their drills. By mid-summer, I was so mightily impressed with my oat plants that I would often steal a few moments to stand out in the field admiring the crop as the breeze rippled across it and the evening sun drenched it in light.

But there was still so much that could go wrong. Harvest time was going to be a problem as the late spring had meant that sowing had been later than usual. We needed almost perfect conditions throughout the year to stand a chance of harvesting in early August. People in this part of the world had become accustomed to the 'Fifth Season' – a period of sustained wet weather that ran through late June into July and which had been a feature of recent years. There were mutterings that it was down to the thirteen full-moons we'd had in each of the last three years and that 'we were due a good summer'. Under wet circumstances our oats would undoubtedly grow well but we needed a dry period in late July and into August to ripen the plants.

The dressed oat seed awaits sowing while I adjust the depth of the seed drills.

I could only pray for the right weather conditions but I could make preparations for the method of harvesting. I would almost certainly be looking to use a reaper-binder and Mr Mudge, a local farmer who had been so generous with his advice and support, had a number of these ingenious contraptions in a near-working state. The reaper-binder leaves the crop cut and bound with only the task of stooking – standing groups of sheaves up against each other – left for us to do in the field. From my experiences on the Victorian farm, I knew the reaper-binder was a heavy and cumbersome machine and required three extremely experienced horses. Although we had three horses on the farm, I was concerned that Jack, who had been a useful cart horse throughout the year, might struggle to work alongside Prince and Tom. If we were to use the horses, some practice runs would be needed but finding the time for this, and the space, would prove difficult.

As much as I hated to admit it, I was going to have to call upon the services of a tractor. Towards the end of the Edwardian period and in the years running into the First World War, immense improvements had been made in the field of internal combustion engines and their motive power around the farm. I'm passionate about horses and would love to have used them for the job but I thought it would be interesting to see just how effective the new machines of the day could be (see page 53).

As Megan kept the horses steady, my job was to check for blockages and ensure an even deposit of seed.

HEDGEROWS AND HEDGEROW MANAGEMENT

The story of the British hedgerow is one that continues to captivate me. Hedgerows are anything but natural entities and were, up until relatively recently, highly managed barriers controlling the movement of stock, marking property and providing precious resources in the form of timber and food. My own area of particular interest is the early medieval period. Old documents often refer to different types of hedge and across the country a whole range of styles can be identified. Frequently, embankments, ditches and walls were incorporated into a hedgerow.

One of my most enjoyable pastimes involves walking the landscape with a detailed map and trying to work out which period of history different hedges originate from. This doesn't just involve counting the number of tree and shrub species – although this is certainly an indicator. It also means looking closely at the physical character of the hedge, the scale of its bank and ditches and how it relates to other linear boundaries in the area. We had a useful array of fields enclosed by what were clearly very old hedgerows and I immediately set about exploring them in some detail. They were majestic constructions comprised of large banks, often stone-faced, with a range of traditional hedgerow shrubs planted into their summits. While many had a single hedge running along the centre of the bank, some places were crowned with a 'double-comb' hedge. This arrangement consisted of two hedges – one either side of the top of the bank – with a narrow path down the middle, and it was these hedgerows on our farm that were in the most desperate need of repair.

I say that hedges are highly managed barriers because a hedge left to its own devices will, after a period of some ten to fifteen years, fall into disrepair and fail to do its job. As certain species grow taller than others, they shut out the sunlight and cause lesser shrubs to die back or grow out in search of light. This, in turn, causes gaps to emerge in the hedge. Badgers, foxes, rabbits and hares pass through these gaps, wearing away the bank and creating larger gaps. They can also cause structural damage to the banks by burrowing into them, while the soil they cast up provides perfect conditions for less desirable hedgerow species such as elder, introduced via bird droppings. Unwanted climbing plants such as honeysuckle, creepers and ivy can all cause the collapse of a hedge.

Many of our hedgerows seemed almost beyond repair. With stock needing to be moved around the farm and crops in need of protection, it was imperative that our hedges were in good order. There was far too much work for me to even contemplate doing alone, so I was quick to call in some help and advice. This came in the form of the Blackdown Hills

ENCLOSURES

Perhaps the most famous period for the laying out of hedges is in the 18th and 19th centuries when, all over the country, the large open fields of medieval villages and their extensive common lands were divided up and 'enclosed' into smaller fields by an act of Parliament. Smaller fields were thought to be easier to farm more productively. Those who did well out of enclosures made profits selling produce to the growing urban centres of industrial Britain. But there can be little doubt that enclosure acts left many poorer folk disenfranchised; families who had for generations survived off a small strip in the open field and a cow kept on the common were reduced to the status of mere labourers.

Enclosure acts had less effect on the south west where, because of its terrain and history, the landscape was already sub-divided into small fields arranged around individual farmsteads rather than villages. Our farm was no exception.

Hedge Association, a charming group of hedge-layers from East Devon, who came armed with a range of tools and a wealth of experience.

I was instantly heartened by their opinion that the hedge banks I had thought beyond salvation could be made stockproof again. Hedge-laying skills include choosing which timber you want to stay in the hedge and which you want to cut out. Then there is the skill of actually 'laying' the living plants horizontally, by making an incision into the trunk close to the base and bending the trunk down. Part of the bark wood is still connected to the root system so that the trunk continues to grow, sending up fresh vertical growth all along its length.

Our hedgerow was mostly blackthorn and hazel, and it is likely it was originally planted with these species as they are both excellent stockproofing shrubs. Yet, a number of invasive species such as ash and sycamore had also colonized the bank along with the odd holly, elder and spindle.

No two hedges are ever the same and a good hedger works with what he's got and, as I found out, the Blackdown Hills hedge-layers were not men to grumble. By the end of the day I thanked the team for all their help, humour and advice and I couldn't wait to get back out there the next morning to crack on with the next stretch of unwieldy shrubs.

MAKING WATTLE HURDLES

Wattling or wattle work is the weaving of pliable rods or 'wands' through a series of upright timbers to create a fence, wall or partition. It is one of the oldest building techniques known to man. Evidence of the practice has been recovered from prehistoric archaeological sites and the principles of the technique occur in other crafts such as textile weaving and basket making.

A wattle hurdle – a moveable panel – is a specific piece of kit with more tangible origins in the sheep farming of the chalk downland. Whereas the sheep-rearing regions of Wales and the Pennines could draw on easily accessible quantities of stone to make boundaries, chalk downland is noticeable for its dearth of stone. Wattle hurdles provided the best way to control movement and manage the flock. As early as the 12th century, coppices of hazel trees were planted and managed for the production of rods.

Traditionally, the hurdle maker would have bought or leased several acres of coppice. Each year he would move to a different area to harvest timber, not revisiting the same stools – trees cut back to ground level to produce multiple stems – for a period of seven to eight years, to allow them to regrow. Coppicing was a winter job. During the summer months sap rises in the hazel and to cut back when this is happening can be damaging, if not fatal, for the tree. Also, timber harvested when full of sap is much more difficult to work.

Colin Risdon of the Blackdown Hills Hedge Association offers me his expert advice on how to lay a hedge.

I'd made fixed wattle fences and wattle and daub walls before, but hurdle making is different: the weave of the hazel has to bind into itself, so that as the hurdle is moved around, ends don't pop out, resulting in the whole job falling apart. Fortunately, local hurdle-maker and green wood-worker Ian Roper was on hand to guide me through the process.

First came the mould – a timber log split down the middle with 10 holes drilled into it. The idea is with the use of a cadgel or mallet, to set the upright 'sails' or 'zails' in the holes. The zails were made from the stouter rods, cut to around three and a half feet (1.08m) in length and sharpened to a point at the bottom. While the zails throughout the hurdle were made from rods split down the middle, the two end zails were made from slightly smaller whole rods, so that the weaving hazel could be twisted round them more easily.

There are two key techniques to successful wattle-hurdle making. First and foremost, you have to get to grips with splitting hazel. This is not an easy task and in more than seven years of trying, it is only in the past few years that I can honestly say that I've learnt how to control the split. The second technique is weaving the hazel. This might sound easy but in fact it requires a good deal of expertize. The rods need to be handled properly to avoid them breaking under the strain.

There is not nearly enough space here to describe the technique. I advise anyone interested to train with a local hurdle maker to appreciate the complexity of the craft and the tricks of the trade. For example, Ian ensured that all the woven ends were neatly trimmed and protruded from the same side of the hurdle. Along with keeping the bark-side of the split hazel all on the same side, this means that as you carry the hurdle, your back comes into contact with the smooth side. Clever. Also, at about two thirds of the way up the hurdle Ian wove some 'twillies' – thinner rods kept 'in the round' (i.e. unsplit) – just short of the central zail before being turned back on themselves. This creates a twilly hole, through which a shepherd can thrust his walking stick or crook before hoisting the whole thing up on his back.

Before removing the finished hurdle from the mould, Ian trimmed all of the loose ends with a 'trimming bill' – a bill-hook with a protruding hatchet blade at its end. This was the only bespoke tool that I didn't have, but Ian was kind enough to give it to me, along with the mould he'd fashioned on our arrival and a cudgel to bash the zails in place. Armed with bill-hook, slasher and cudgel, I could set about making my own wattle hurdles.

Local wattle-hurdle maker Ian Roper and my wattle hurdle.

RICK BUILDING

A rick is a method of storing hay in a field by piling it up and putting a basic thatch on it (often using materials such as water reed or bracken) to protect it from the elements. Hay can then be cut from it and fed to the stock as and when it is needed, throughout winter, until the grass begins to grow again in the spring. Putting the hay in a field saves on barn space and ensures that if the hay spontaneously combusts (as it will do if it is stacked wet and begins to ferment and then oxygen gets to it), you don't have to file a claim on your farm buildings insurance.

The shape of a hay rick will depend upon its location. If it is on open flat land, then the rick will usually be round to prevent the wind taking out a corner; if it is in a more sheltered spot, such as a valley, then chances are it will be more shed shaped. The size of a rick will obviously depend on the amount of hay that the fields have produced; some old pictures depict huge hay ricks larger than many historic farm buildings. *The Book of the Farm* (a manual written by Henry Stephen in 1855) features horse-powered elevators, cranes and giant frames that suspend canvas sheets in case the weather turns nasty mid-build – all of which were needed to build huge ricks. However, rick building began to decline with the increasing popularity of large Dutch barns and baling technology, which began in the late 19th century in the form of hay presses and culminated in the 1940s with mobile balers.

Therefore it wasn't surprising that when we began to search for an expert – someone who had perhaps done this process or seen it done as a child – we were told that we should look in the graveyard. I'm sure there are many still alive who will lay claim to rick building, but Alex and I were on our own. Our first decision (which was to prove our most astute) was to build the rick in the stack yard on top of the staddle stones. These are mushroom-shaped stones that have been in use for millennia to prevent rodents from getting at the grain stacked on top of them. There is no grain in hay but putting it on top of these stones allows air to get all around the hay to dry it and keeps the hay off the ground, which over winter was often awash with water – such is the nature of the Tamar Valley. It also gave the poultry somewhere to shelter and because the stack yard is stock proof it meant that we didn't have to fence off the rick to prevent the cows, sheep and horses from eating it.

ALEX'S DIARY

It was mid-winter and with great trepidation Peter and I cut into our hay rick. A hundred years ago, this was the point that could make or break a farmer. As we peeled back the first straw mat, my foot squelched on the sodden hay. Small mushrooms growing from the crop instantly informed me of the damp, humid and warm conditions – disaster! I was plunged into quiet despair but Peter, somewhat more optimistically, peeled back more of the mats. It soon became clear that the part I had exposed represented only a tiny part of the rick where water had 'gullied': luckily the majority of the rick was bone dry. My spirits lifted and, with a newly sharpened rick-knife, we set about cutting out swathes of fodder. What touched me most about the whole project was that even by the Edwardian period, rick building was dying out. We had nose-dived blindly into a truly bygone craft and had been successful. We were both chuffed to bits and often stopped to stare at the clean-cut section through the rick and admire the dry hay.

We began our rick by collecting timber and arranging it on the staddle stones in a herringbone formation to provide a frame upon which to place the hay. We then began pitching the hay on top, building up the edges first and then filling in the middle – much like the techniques we had used each time stack a dray with a crop to bring it down from the field. The key was to make sure the loads of hay overlapped, the layers went on evenly and that we trod around the rick as we constructed it, compacting down the material. As the height of the rick increased so did its springiness and the amount it rocked back and forth, and this was a good indication of the integrity of the structure.

It was hard and thirsty work but we got there in the end knowing that we had peace of mind for the winter and being able to stand back and admire a soundly built hay rick.

When it came to using the hay we managed to find a couple of rick knives (imagine giant kitchen knives) on site that looked as if they could do the job. However, we struggled, despite oiling and sharpening them. Then a local man Stephen Veal gave me his and it just ate through the hay like a bow saw through balsa wood. We cut out the rick in blocks so that we had hay platforms to work off of. All the pieces of hay had been laid horizontally, so that each face that we cut was made up of thousands of cross-sections of circular stalk ends. It was amazing to see how compact it was and how dry the hay was.

For me this was a real success, a true insight into a farming process that hasn't been practised en masse in this country since probably the Edwardian period and it emphasized the reason why we take on such projects.

THATCHING A HAY RICK

Deep down, one of my main reasons for wanting to build a rick was that it gave me the opportunity to try my hand at thatching. I love thatching. I've thatched barns, privies, hovels and even one of my chicken sheds, but I'd never thatched a hay rick.

Where the thatching of ricks departs from the thatching of cottages is in the lifespan of the thatch. Essentially, a rick needs to last no longer than the winter and as a consequence a whole range of materials and styles are known throughout the British Isles. One of my favourite illustrations from Henry Stephens' 1855 edition of *The Book of the Farm* is of the many methods in which straw- or heather-made ropes are used to fasten down a thatch on hay ricks in Scotland. This is truly an ancient method and one some experts believe may have been used in the thatching of prehistoric roundhouses.

Equally, the materials for thatching could differ. In wheat-growing areas a surplus of long straw after harvest time would provide the perfect material. The best stuff would be reserved for thatching dwellings, while the shorter and rougher material was ideal for the thatching of ricks. In areas where cereal straw was in short supply, rushes, reeds, bracken and marram grass were substituted.

Peter rolls out our newly made thatch mats while I stake them down.
OPPOSITE Keith, Alex, Bill and Peter assess the total roof space of our hay rick and decide on the quantity of thatch matting needed to weatherproof it.

Having built our rick, Peter and I needed to let it settle for a fortnight before attempting to apply a thatch. Knowing that this type of project would really excite him, I contacted Keith Payne, who originally taught me how to thatch. Keith was wildly enthusiastic and turned up with a surprise in store.

The 'Thatch Mat-maker'

One of the things I'd read about in *The Book of the Farm* was a 'thatch mat-maker' and I'd mentioned this ingenious contraption to Keith. In essence, this giant sewing machine stitches together loose, long, wheat straw to make a roll of matting. This can then be carried to the rick, rolled out and sparred (pegged) into the hay. It turned out that Keith had a mate, Bill Liversage, who owned a mat-maker. When Keith turned up he brought with him a whole bunch of loose straw, Bill and his machine, and an enormous flagon of cider. Once the machine had been wrestled into place, Peter provided the hand-crank power, Bill fed the loose straw in one side and I eased the newly stitched mats out the other side. In the meantime Keith had been splitting hazel to make the spars to fasten the mats, and in no time we were ready to fix up our thatch.

There were a number of advantages to using mats. First, when we needed to use our hay, we could just peel the mats back and cut in. When the hay was finished, the mats could be rolled up and stored for next year's rick or for other jobs around the farm – such as windbreaks during lambing or covers for a potato clamp. Perhaps the most keenly felt advantage though, was the inordinate amount of time we had saved. Although I'd been desperately keen to have a go at some more thatching, there was so much more work that needed my attention. A job that might have taken three to four days had taken us less than an afternoon. So we had more time to enjoy Bill's cider, which turned out to be equally as effective as his thatch mat-maker.

Bill Liversage and his thatch mat-maker.

SILAGE

I don't know how we manage it, but in the wettest summers we make hay and now in one of the driest summers that we have had for a while, we decide to make silage at Morwellham Farm. Silage is thought to have been introduced into Britain from continental Europe during the 1880s, as a method of making winter fodder in wet summers. At the start of the 20th century, when farming became more scientific and the government began to get more involved, silage was championed as a fodder crop. It has increased in popularity ever since, with production spiking during both world wars and really taking off in the 1970s.

To make our silage we set about cutting our grass with a single horse-drawn finger-bar mower (silage is best made from grass that has been chopped into short lengths). Then we collected up the grass using a hay sweep. This is a very old piece of kit made largely out of wood and looks like a large comb with handles. It is dragged along the ground and collects up all the loose hay. However, hay generally 'sits up' and is quite light, whereas the grass we had cut laid flat on the ground and was wet and matted. Initially I didn't think the hay sweep would be an effective way of collecting the grass, but we persisted and eventually it began to do its job, especially once the grass had been swept over and rolled up into wads – on subsequent sweeps these could be picked up quite easily. In addition, this process helped pack the grass tightly together and force out excess air, which is essential for good silage making.

Once the sweep was full of grass we emptied it into a pile, close to where we proposed to build our silage clamp. To do this, you simply lift the handles of the hay sweep. The sweep is pulled by a single horse on long chains (at least 12 feet/3.6m) and when the handles are lifted the whole piece of kit does a somersault and the contents are deposited on the ground. The long chains are to prevent the apparatus from striking the horse. This was a very quick and effective way of collecting our grass.

Building the Clamp

Our turn of the century agricultural leaflets recommended the use of a clamp to store the silage. We piled up our grass much in the same way that we built our rick, starting at the edges and filling in the middle, trampling it as we went to ensure maximum compaction. We sculpted an elongated pile that was approximately 5 feet (1.5m) wide at the base, 3 feet (90cm) high (with tapering sides) and 40 feet (12m) in length. When we were happy with the compressed shape we covered it with the straw mats that had

WHAT IS SILAGE?

Silage is made from grass crops, including cereals, and needs moisture content to essentially pickle it, making it an ideal way to deal with would-be hay meadows during a wet summer or in a notoriously wet area. To make silage, you cut the grass and collect it up while it is still green. The grass is then stored in such a way that oxygen can't get to it and it begins to undergo anaerobic fermentation. (Translated from Greek, anaerobic literally means 'life without air'.) This method helps the grass retain more of its nutrients and avoids having to supplement hay with a root crop (such as mangel-wurzels). Silage is generally fed to ruminants such as sheep and cows.

PRESERVING CROPS

A clamp is a method of preserving crops such as roots in a field for many months. The crop tends to be arranged in a long, thin alignment and covered with straw and soil. The clamp can then be accessed from one end, as and when the fodder is needed, without disturbing the remainder of the contents. Clamps date from at least medieval times and are a great way of storing winter fodder crops without taking up barn space.

previously been used on top of the rick and used spars to hold them in place. These mats act as both a structural support and as a barrier, to keep the earth that we piled on top away from the silage.

It is thought to be in Iceland in 1922 when the first links were made between silage and listeria monocytogens, which cause listeriosis. (This is also known as the circling disease or silage sickness in livestock.) Listeriosis in silage mainly occurs when the silage is poorly made, with air gaps or when it is contaminated with soil. Therefore straw mats were an essential element to keep the silage clean.

To cover the clamp with soil we dug a horse-shoe shaped trench around the pile. It was open at the lower end and would protect the silage from ground water. We cut turfs from this trench that were like large bricks bound together by root matter. We used these to build a turf wall at least 5 courses high, which came up close to the top of the clamp. A clamp should be covered with 8 inches (20cm) of soil and by building a turf wall we succeeded in doing this as well as saving ourselves a lot of time. On the top of the clamp we placed buckets of soil that we had dug out of our trench. We then patted this soil down, stood back and admired our work.

After building a silage clamp it takes about 48 hours for the pickling process to begin. The silage should be ready after two weeks, but of course it will be left in the clamp until it is needed as a fodder crop over winter. One of the drawbacks of making silage is that it produces a lot of nitric acid and the clamp should be sited somewhere where it won't interfere with the water system. This was a job well done on the Edwardian farm.

The completed silage clamp in the foreground with the potato (right) and oat (left) crops in the background.

THE COMING OF THE TRACTOR

One of our major winter tasks was ploughing Cottage Field for sowing oats in early spring. I was hoping to get most of it done before it got too cold and the ground froze over, but what with problems in finding a plough, and taking time to get used to the horses, things got delayed. By mid-December, we had ploughed a mere strip of the field and the frosts were starting to set in hard. At this point the production team asked how I would feel about bringing a tractor in to do the work. At first, I balked at the idea. In any event, I was quick to point out, tractors from the Edwardian period were so rare and valuable it was unimaginable that anyone would let such a price-less vintage piece of equipment near our muddy fields. Not so, it transpired. The production team had done their homework and found the oldest working tractor in Britain and they were extremely keen to give us Edwardian farmers a taste of what it was capable of.

The 1903 Royal Agricultural Show silver-medal-winning Ivel Agricultural Motor made a fantastic noise. The water-cooled engine hissed and popped and the beautifully polished frame clanked and clunked as it entered the field. Behind, it towed a two-furrow ride-on plough and I secretly looked forward to the enormous amount of work a successful demonstration might save me.

Teething Problems

We tried to imagine we were early 20th-century farmers and prepared ourselves to be wowed by the introduction of a machine that would change the face of farming for ever. Sadly it wasn't the case. Teething problems with the plough meant that the first few furrows were shallow and very roughly laid down. In an attempt to increase the depths of the plough shares, the greater drag caused the Ivel to lose traction. The large but thin rear wheels, despite their treads, immediately started spinning and spraying mud. Increasing the power only made things worse – the field was getting badly chewed up and we decided to abandon the ploughing demonstration. I didn't want to gloat. But I did.

The next demonstration took place in the threshing barn where we had set up a corn mill to be belt-driven by a pulley wheel on the Ivel. Once the large belt was attached to both tractor and mill and the engine fired up, I saw the benefits of this compact, powerful machine.

THE IVEL TRACTOR

The Ivel Agricultural Motor was the brainchild of Dan Albone and, patented in 1902, represented the first successful attempt to introduce the internal combustion tractor to British farms. Steam-powered traction engines had been around since the 1890s and Hornsby-Ackroyd had, in 1897, produced an oil-powered machine but both these were heavy and slow implements and more of an expensive hindrance to most farmers. Albone's genius was to create a much lighter frame, on to which he mounted an 8 horse-power, 2 cylinder engine and it may very well have been his earlier successes as a bicycle and tricycle manufacturer that gave him the vision to think small-but-strong. The resultant Ivel (named after the River Ivel in his native Bedfordshire) was both light and powerful, with a load capacity of 2.5 tonnes and a top speed of 5mph (8km/h).

The corn-milling demonstration was an unbridled success and as fast as I could feed the grain into the mill, it was being crushed down and spat out into a bucket below.

In many ways, the day's demonstration had been a true reflection of where the tractor stood in the eyes of Edwardian farmers. Stationary petrol engines had become almost commonplace on Edwardian farms with around 13,000 recorded in the 1908 census of agricultural production, and from our demonstration it was clear to see why. But tractors – or rather 'agricultural motors' as they were called back in the day – were subject to overheating, slow and cumbersome to handle, and still not quite up to heavy-duty field-work. A pair of working shire horses were safer, quicker and more reliable.

Nonetheless, the writing was on the wall for the shire horse. H. P. Saunderson of Bedfordshire and his Saunderson Tractor and Implement Company was at one point the biggest tractor manufacturer outside the United States, and in 1910 had moved on

RIGHT As a static engine there could be little doubt that the power output of the Ivel surpassed, more effectively, all modes of power that had come before. A belt from the Ivel agricultural motor powers our corn-milling machine to great effect.
OVERLEAF In the wet and sodden conditions in our arable field the Ivel Agricultural Motor lost traction quickly and struggled to plough.

from the three-wheeled tractor frame to a more stable four-wheeled model. The high point for Saunderson's company, founded in 1890, was the beginning of the First World War, when a labour shortage generated demand for motorized mechanical power. But British manufacturers couldn't keep up with the American production line and when Henry Ford, the man who gave America the Model T, turned his attention to agriculture, he very swiftly flooded the market with the popular Fordson tractor. Tractors did the majority of all farm work during the war, as more and more horses were commandeered to haul guns, ammunition and supplies on the frontline.

For me, it had been a truly insightful day and the Ivel represented a remarkable piece of agricultural history. But the next day I knew I would be harnessing up Prince and Tom and getting back to business as usual – ploughing as it had been done for centuries before.

THE ANIMALS

Perusing the supermarkets it is very easy to dismiss animals as cuts of meat. Working a small historical farm those very same beasts take on a whole new dimension. It may sound contrived but each sheep, cow, pony, chicken, duck, turkey and goat has a very distinct personality that you get to know over the course of the year. Our job is to look after them as best we can, provide for them and protect them, giving them the best life possible. In return, they give themselves.

One of our Red Ruby calves; over the year our herd grew by 50%.

RED RUBY CATTLE

For the series *Edwardian Farm* our choice of cattle for this area of England came down to one of two breeds: either South Devons, which are a yellowish brown colour, or North Devons known as Red Ruby cattle (or occasionally Ruby Reds). We opted to go for the North Devon Red Ruby cattle, a breed that was favoured by the Duke of Bedford.

As you might expect of a breed that has spent the best part of two millennia living in the often inhospitable conditions of the Devon moors, the Red Ruby are very hardy creatures. They have thick skins and are good walkers and foragers, with the ability to survive on grass alone. This also makes them a very good breed to fit into modern countryside initiatives such as the Environmental Stewardship Scheme. They are also very docile cattle, quick to mature, and have a reputation for calving easily.

Red Ruby can tolerate a range of extreme temperatures and subsequently they have been exported around the world. The breed was taken to America by the Pilgrim Fathers from Plymouth and there are herds in Australia and South America, which may well be associated with the miners from the Tamar Valley and surrounding areas who emigrated there. Initially the cattle were a dual-purpose breed: however, in the past 50 years they have been bred for their meat – although in the United States there are still 'milking Devons' retaining those original characteristics.

Their historic pedigree, along with breed qualities, made the North Devons the ideal candidates for us to try our hands at cattle rearing on the Edwardian Farm. We had some experience with cows before – looking after a small herd of Welsh Mountain Blacks (in *Tales from the Green Valley*) and keeping a couple of Shorthorns (in *Victorian Farm*). In both instances our cows had come to us in calf. This time we were going to send in the bull: enter stage left, Shillamill King David II. Our bull came to us from David Hutchinson and, like the herd he was about to join, he was fairly laid back. The cattle all bonded instantly and King David was generally to be seen in the middle of the group each time we checked them. However, he did not become the leader of the pack – that role was retained by the matriarch cow.

When we first took charge of the herd, they moved smartly away when they saw us coming but, over time, as we fed them and stood in the field with them, we reduced this to the point where we could be within the herd – a bond that was ultimately demonstrated by our cattle drove.

Cattle Drove

Britain has an ancient network of drove roads to enable the movement of livestock, especially cattle, from one end of the island to another. High, tight, well-laid hedges and dry-stone walls aided this stock migration. Many towns or villages have a 'holloway', a track that has been well worn by countless hooves over the years and is often on a slope. In Devon, Dartmoor is famous for its white and red 'tides' as farmers moved their sheep or cattle to fresh pasture or to market.

Therefore, when our pasture became thin due to the harsh winter and our hay supplies began to dwindle, it made sense to Alex and me to drive our cattle the three miles through the woods along the old abandoned railways from Morwellham to a farm further up the valley where we had been offered some new pastures. However, neither of us had done this before and the cattle had never been driven such a long distance – let alone on a route that had a sheer drop on one side and mine shafts on the other. We were totally reliant on the relationship we had built up over winter with them, when we would feed them and check them each day and spend a bit of time with them.

Prior to the drove we did as much preparation as possible. We walked the route noting any pitfalls (literally) and points where other tracks broke off, and any areas

ORIGINS OF THE RED RUBY

They are believed to have their origins in pre-Roman Britain and said to be descended from Bos longifrons (small Iron Age cattle) and Urus or Bos primigenius – an ancestor of domestic cattle in Europe. There have been suggestions that breed stock was brought over from North Africa and the Middle East by the Phoenicians (whose civilization was at its peak from 1200–800BC) while trading for tin in Cornwall – however, this point of view is largely disputed.

Early improvers of the breed were Francis Quartly and his brothers William and Henry of Great Champson Farm, on the edge of the village of Molland on the south-west corner of Exmoor. John Tanner Davy and his brother William, of Rose Ash, started the herd back in 1851 and the Red Ruby breed society was founded in 1884.

NEAT'S FOOT OIL

Working with horses on a daily basis means the leather harness is exposed to the elements, the soil and the sweat from the horses (and of course oneself). To maintain the condition of the leather it needs to be cleaned and, if it is traditional oak-tanned leather, it will need to be treated (to avoid the tanning process reversing). One of the best substances for looking after leather is neat's foot oil, which is often a component of leather care products.

Neat is the Old English word for cattle – a clue that it is a complementary product to cow-hide leather. The oil occurs in bovine feet and shins (but not the hoofs) and stays fluid at a low temperature. In winter when the ground is frozen, the oil prevents the lower legs of the cattle from freezing – their legs are very thin (in relation to the rest of the body) with little insulating tissue. Neat's foot oil is easily absorbed when applied to the leather, filling up the cracks and gaps and generally keeping the harsh working conditions at bay. It also darkens the leather.

To extract the oil I procured 12 lower bullock legs free from the abattoir. With a limited oil extraction plan, based on even less knowledge, I set to work. Stage one was to light the copper and bring the water to the boil. Stage two was to immerse the legs and stage three was to maintain a rolling boil. This is one of the strangest things that I have done! Initially the legs fitted in the copper, but as they heated up and essentially 'cooked' they swelled, so that the hoofs started to push up the lid. Not an activity for late at night. The smell was one of beef stew mixed with beef poo, as I had not cleaned the hoofs first. One reaps what one sows!

Stage four was 'proof of the pudding'. I had pinched a number of sweet jars from the shop and with pokers placed in them to conduct the heat and prevent them cracking, I ladled out the brown water into the jars. The oil was visible on the top of the mixture and I left it in the jars to settle and cool. Later I skimmed the oil off the top and put it into a container for use up at the tack room. The oil was a vibrant golden colour and was viscous but fluid – very similar to a thick olive oil.

Cleaning the copper was not a fun process. I had anticipated a certain amount of gelatine to be present in the legs but little did I know that the 'liquid' in the jars and the remaining 'liquid' in the copper would be brown jelly with hair and bone fragments. However, it was well worth it, as the oil that I applied to the leather was fantastic quality, and a little went a long way.

This surreal sight is the bovine feet prior to boiling.

where we could hold the herd – such as at the stream, where they would probably drink. We moved the herd to the field nearest the farm and didn't feed them their usual hay ration the day before the drove, gambling that they would follow us in the hope of food. We also separated out two of the cows due to give birth (as we planned to milk them to make clotted cream) and two of the calves to keep them company, but we made sure we left the matriarch cow in the herd, whom every other beast (including the bull) would follow. However, it was going to be a case of learning on the job and responding to the herd as appropriate.

There was a connection of trust and general understanding.

We decided to start the drove at first light while the mist still shrouded the valley and the sun was yet to break over the horizon. Alex and I armed ourselves with stout sticks to increase our reach should we need to. With everyone assembled and keen to get going and all relevant gates open (or shut), it was time to get the cows. Would they take to the drove like their ancestors or would we be spending most of the morning trying to find our bovine friends in the dense woods of the Tamar Valley?

The cows came to the call of our voices and we led them through the familiar territory of the farmyard toward the unknown land of the incline railway path. So far so good, as we crossed the first danger zone: the crossroads that lead off either uphill to the canal or downhill. It was starting to look as if all our fears had been unfounded, when our cows suddenly took a left and headed downhill towards the river. Alex raced through the trees and brambles, cutting the herd off and diverting them back towards

Our herd moving as one to the sound of our voices as we drove them through the hills.

the path, while I regained a sight line with the matriarch and continued calling. We had them back under control, back on track and going where we wanted, when again our cows took a left and headed downhill . . . déjà vu. The third time they carried on, following my voice, heading uphill towards Morwell Rocks.

When moving cattle I try and use two calls: a 'come on, come on cows, come on' accompanied with minimal body movement when I want them to follow me; and a linguistically sharp 'hut hut hut' coupled with feet stamping and arm waving, when I want them to go away from me. On the drove I was in front the whole time and my call became melodic, rhythmic and continuous, to the point of being hypnotic, echoing throughout the peaceful valley below. The herd moved at quite a pace and it helped to go a little way ahead of them when coming up to junctions, so as not to form a barrier and divert them down the wrong track. Equally it was important that no individual cow got past me, so if they were getting too close or going too quickly, I would hold my stick horizontally above my head and with straight arms draw it down and up, which would subdue them.

This was one of the most amazing things that I have done with animals. When we reached the other side of the woods and Sarah Birt's farm, I had an amazing feeling of euphoria. The morning light had been magical when it backlit the herd and steam and dust were rising from them, shortly after we had stopped at our little stream for a drink. Only once on the journey had my heart leapt into my mouth, when I thought for a split second that the cows had managed to get on the wrong side of the barrier at Morwell Rocks. Finally, when they were let loose in the field of lush green grass, it was as if they had never met us; they were a herd once again and we were no longer food providers, we were just predators to be wary of.

One of our cows enjoying a welcome mouthful of hay.

GOATS

The native Old English goat is a hardy beast, but not in truth a terribly useful creature on an Edwardian farm. Her milk yield is low, she produces no wool and her meat is very lean. Large herds of goats did not make economic sense. Sheep and cattle were much more efficient in turning grass into a saleable crop. But for some people the goat was a useful if small-scale addition to the farmyard.

During the latter half of Victoria's reign dairy farming had changed substantially. With the coming of the railways it had become possible to send fresh milk to markets far away. Daily collections of milk cans were taken to large towns and cities where, still only hours out of the cow, the fresh milk could be sold for good money. Dairy farms expanded to meet this new market wherever there were good railway collections. And in 1880 there were four times as many railway stations in Britain as there are now, many of them deliberately in rural areas to provide freight transport for agricultural business.

But while the country was producing more milk than it had ever done in the past, very little of it remained in the countryside for the rural population to drink. The diet of rural children was a cause for concern and goats were the answer.

Getting the goats to stand for milking could be something of a challenge.

Smelly, boisterous, will eat anything – no not Alex, Ruth or I, but our goats. It quickly became apparent that, to look after these beasts, bribery and corruption in the form of tasty leaves would be necessary. We also had to get to grips with milking. Unlike a cow or sheep, the action isn't a straight down stroke of the teats: instead you push up, so that the milk fills the teat. Then using your forefinger and thumb, you form a seal before the the milk is squeezed out. Doing this on a daily basis I found it a lot easier to reach around from behind the goat, rather than just milking from one side – a technique that I was assured was employed in Tasmania in Australia (where there is also a Tamar Valley complete with the city of Launceston, which is fairly close to Devonport and Torquay). After each milking we made sure we had removed all the milk from the teats, to reduce the chances of mastitis occurring. It is surprising how much milk the goats produce and how well they can strip a patch of land. As long as you can keep them away from the garden, they will maintain your milk stocks with relative ease.

Grazing Pasture

Goats do not need good pasture. They do equally well, maybe even better, grazing on scruffy bits of wasteland, along hedgerows and so forth. Being small, a goat can be tethered on a narrow patch of roadside verge and moved on daily, eating not just grass but more or less any green thing. This means a goat can be an extra farm animal, not just a replacement.

We didn't have any cattle in milk for the first half of the year, so it seemed like a good idea to try out a couple of goats. There was no shortage of overgrown land to graze them on, and we hoped they might prove useful in helping us clear all the vegetation from the tumble-down walls of the market-garden terraces – as well as keep us supplied with a little winter milk.

Sheep farming on the British Isles would have been impossible without the services of man's best friend. Without Kenny Watson and his charming collies, Wag and Roy, driving the sheep across the open wastes of Dartmoor would have been impossible. (Kenny is a local Dartmoor farmer born and bred. He keeps cattle and sheep, but is best known for his excellent skills at breeding and training sheep dogs.)

SHEEP BREEDS OF SOUTH WEST BRITAIN

Where would the British landscape be without a flock of sheep scattered across the distant hillsides? Britain is, and always has been, famous for its sheep. Although countries like China, Russia and Australia have by far and away the largest populations of sheep, it is Britain that has the greatest variety of breeds. This variation is perhaps due to our landscape, where there is a fantastic mix between upland hills and lowland pastures.

Since the dawn of domestication there have been sheep to hand, raised in equal measure for their milk, meat and wool. DNA analysis of bones found on archaeological sites dated to the Neolithic period has determined our earliest sheep to have been closely related to the Mouflon sheep of central Asia; while the Soay sheep, our oldest surviving species and resident of the remote island of St Kilda, is believed to have its origins in the Bronze Age. The Romans introduced what today are called 'longwools' and nearly two thousand years of breeding between these and the hardier native sheep has created the rich variety of breeds we know today.

A Devon longwool (left) and a Whitefaced Dartmoor (right) share a bucket of milled oats.

The Importance of the Wool Industry

Back in the medieval period the economy was founded on the wool industry. The emergence of Britain as a global power was founded very much on the wealth it accrued through the sale of both the raw material and the cloth made from it. Devon and the south west played a key role in this industry. You need look no further than the plethora of Devon place names with a wool prefix (Wooladon, Woolacombe, Woolfardisworthy, Wolborough) to realize the importance of the wool industry.

With the Industrial Revolution and the growth of sheep farming abroad, the wool trade collapsed and an industry that had been key to rural life in Devon fell into decline. There was still a local demand for wool, but sheep began to be reared more for meat than for wool alone. This lead to a system of sheep rearing in Britain that is almost unique in the world, where cross-breeding actually maintains the integrity of old pedigree breeds – while at the same time producing excellent crossbreeds or half-bred sheep for meat.

BELOW Peter and Alex go head to head in a hand-shearing contest. OVERLEAF From right to left – Peter, Barry Quick, George Mudge, Colin Pearse, Andrew Mudge and Alex.

The concept is quite simple. Hardy and resilient hill breeds are kept up on the hills where they can survive off relatively poor pasture. Once they are three or four years old, they are 'drafted' down to lower pastures where they are crossed with larger longwool breeds. The resultant lamb is termed a 'half-breed' or 'mule'. The ram-lambs of this generation are sold off for meat and the ewes are crossed with a pure-bred ram from a good meat breed – such as a Suffolk or a Texel. The longwool breeds nearly always produce twins and regularly triplets, so their high productivity is combined with the excellent forage conversion capacity of the hill breed. Stratified breeding, practised all over the country from the end of the 18th century, ensures that the lineage of the respective breeds stays pure, but that also the best qualities are gained from each breed and meat productivity is maximized.

ALEX'S DIARY

We wanted to recreate that special time of year for the Dartmoor sheep farmer when the sheep are shorn and driven up on to the wilder stretches of the moor to enjoy the summer grazing. The purpose of driving the sheep up on to the moor was to give the farm a well-deserved break, to make use of the summer growth of wild grasses, gorse and heather as fodder and to 'fitten up' the sheep – to get them lean and hardy in preparation for going to the ram in October.

Firstly, however, they would need to be shorn of their fleeces which, many moons ago, were the most important part of the animal. Now, with wool commanding such feeble prices at market, it has become something of a burden. If left unshorn, a sheep may very well suffer from exhaustion in the summer heat, so it was imperative that we remove the coat that had served them so well in the cold winter months.

Peter and I had tried shearing by hand before and had found it to be a laborious process. So we were both extremely anxious about shearing the flock of White-face Dartmoor sheep. Mercifully for us, we weren't going to be alone as father and son combo George and Andrew Mudge were on hand to take us through the process. And my word,

ALEX'S DIARY

were these boys quick. In fact, they are so quick at shearing sheep by hand that they have represented England at the World Shearing Championships.

I was lucky if I could, within 4 minutes, get the sheep to sit still enough to shear, yet in that time, both George and Andrew would have sheared a sheep each! Peter and I watched in awe as they sheared their way through the flock at high speed and we were justifiably nervous when we were handed the shears to compete against ourselves.

However, with Andrew coaching on my shoulder and George on Peter's, we felt something of the competitive streak that had driven both these shearing champs to compete at the highest level. Peter thrashed me by a good minute and a half and had certainly removed the fleece from the animal quicker than I. I took some consolation from the fact that my fleece was at least in one piece. . .

TRADITIONAL SHEEP BREEDS
OF THE SOUTH WEST

DEVON LONGWOOL

The Devon Longwool is found throughout the south west of England. It is a polled breed, which means the horns have selectively been bred out. The sheep have white faces and black nostrils, and their thick wool covers much of their face and legs. Their fleece is long, curly and strong, and today their wool is valued for making rugs and carpets. Back in the medieval and late-medieval periods and well into the 19th century their wool was used to produce serge (a coarse woollen cloth), tweeds and braids. The Devon Longwool is a grassland sheep bred for the more luscious pasture of the valley bottoms. The breed suffered in the later Victorian period when an increase in dairy farming meant more pasture was dedicated to cows.

Ram 200–300 lb (90–140 kg)
Ewe 165–190 lb (75–85 kg)
Fleece 11–15 lb (5–7 kg)

EXMOOR HORN

The Exmoor Horn is a descendent of the horned sheep that roamed Exmoor in days of yore. Hill breeds are crucial to the ecology of the moors in this region: like their famous friends the Exmoor ponies, they maintain the pasture in a grazed state and prevent it reverting to shrub and ultimately forest. The breed has fine-quality wool, which is rare for a hill breed. It was once the most sought-after meat for the London restaurant trade in the 19th century. The Exmoor Horn is an ideal crossing ewe as it is both a good mother and milker. The Exmoor mule – the product of cross-breeding with the Bluefaced Leicester – yields fine lamb for the modern market.

Ram 175–245 lb (80–110 kg)
Ewe 130–190 lb (60–85 kg)
Fleece 4½ – 5½ lb (2–2.5 kg)

GREYFACE DARTMOOR

A product of improvements carried out in the 19th century, the Greyface Dartmoor owes its origins to cross-breeding the native Dartmoor sheep with the local longwools. The resultant breed has a distinctive fleece of long curly wool (classified as lustre longwool), which covers both the head and legs. The greyface tag comes from the grey or black spots that mottle their white faces. It is also known as the improved Dartmoor.

This polled breed was bred for its wool which, like the Devon Longwool, was sought after for coarse fabrics such as serge, carpets and rugs. The ewes are good milkers and as far as productivity goes, a flock should return 140 per cent success rate at lambing time – that's 1.4 lambs per ewe on average. The Dartmoor Sheep Breeders' Association was started in 1909 to standardize, promote and develop the breed – which had, by then, become well established in many areas around the foothills of Dartmoor.

Ram 130–200 lb (60–90 kg)
Ewe 110–145 lb (50–65 kg)
Fleece 9½ –13 lb (4–6 kg)

WHITEFACE DARTMOOR

Whiteface Dartmoor sheep have white heads and faces, with both the head and legs free of wool. Their nostrils are black and their ears are short and thick – always a good indicator of hardiness in sheep. The wool they produce is moderately greasy and has a fairly strong curl. They are a hardy breed and capable of producing both a good lamb and fleece through wet weather and hard winters. The Whiteface Dartmoor is thought to be one of England's oldest breeds and was once much more widespread but has been pushed back by centuries of enclosure to the extremities of Dartmoor, where it has been forced to develop on some of the poorest pasture, grazing at heights of between 500–2,000 feet (150–600m) above sea level.

As a breed, it has proved successful in more recent times as a 'cross' on the Devon Longwool and, further afield, on the Welsh Mountain and the Suffolk. The heyday of this ancient breed was in the 17th and 18th centuries – sadly today only a small number of the quarter of a million sheep on Dartmoor are Whiteface Dartmoors.

Ram 130–200 lb (60–90 kg)
Ewe 110–130 lb (50–60 kg)
Fleece 6½ – 11 lb (3–5 kg)

Our large black pigs had insatiable appetites and whined constantly for food.

THE LARGE BLACK PIG

The Large Black is Britain's only black pig. It is a docile and hardy pedigree breed renowned for succulent pork and excellent cured bacon and originates from the Old English hog of the 16th and 17th centuries. By the late Victorian period the breed had become localized in two extremes of the British Isles – in the south west and in East Anglia. In the 19th century steps were taken to bring the two populations together, with breeding programmes that drew from stock in both locations. By the time the Large Black Pig Society was established in 1889 the breed was known for being vigorous, with the sows producing ample milk and able to raise a sizeable litter of piglets on unsophisticated rations.

Thanks to the society, the Large Black became a popular breed in Edwardian times and was crossed with large white pigs from various parts of Britain to bring out the best qualities of each breed. The breed found success in the show ring, culminating in one sow fetching 700 guineas after becoming overall champion at the 1919 Supreme Championship of the Smithfield Show. In the same year, Large Black pigs outnumbered all other breeds at the Royal Show, demonstrating just how popular this breed had become.

When the Duke of Bedford set out to provide accommodation for the miners who worked in and around Morwellham Quay, the cottages were designed with an intriguing arrangement of outhouses at the rear of the buildings. These consisted of a privy with a pigsty to one side, which were both connected to a small yard accessible from the outside. This set-up harks back to the days when it was an indication of social status whether or not you had the facilities to fatten your own pig. The sties were designed mainly for weaner pigs, which could be purchased from the local farmer as piglets only just weaned from their mother's milk and fattened on a diet of typical Victorian backyard waste. In the interests of 'bio-security' we, of course, ensured that our pigs had a healthy and varied diet, along with a regular run in the pig paddock where they could grub around for nature's treats. There is something extremely satisfying to the animal-welfare-conscious meat eater to see pigs happily basking in a sunny field having dined on a bucket of carrot tops, turnip peelings and cabbage hearts.

 ## ANIMAL HOSPITAL

Over the year the end stall of the stable in our range of farm buildings has served as an observation pen for a variety of animal patients. One visitor was a crop-bound cockerel. He had obviously been very greedy, eating a lot of seed – much of which he had stolen from a goat mixture that contained un-milled oat grains. A chicken stores its food in a pouch in the gullet known as a crop, but this cockerel had

eaten far too much and probably had not been drinking enough water, so it was unable to empty its crop naturally. The situation was exacerbated by the fact that the birds blocked crop stops any food getting into the stomach, so it gets hungrier and attempts to eat more, compounding the blockage in the crop.

Alex dresses a head wound on a Dorking hen, the likely result of a failed fox attack.

Our solution was to pour luke-warm water mixed with a little oil (such as olive oil) into the crop and massage the contents gently. We then flipped the chicken upside down and the water came back out, bringing with it some of the contents that were causing the problem. We repeated this process a few times a day for a couple of days and I am pleased to report that he recovered and is still doing well.

Another visitor was one of our geese. She had been caught by the fox which had left a tell-tale puncture wound deep in her back but somehow she had survived. However, she was unresponsive and couldn't lift head nor wing; when she tried to walk, she would more often than not go backwards before collapsing in a heap. We feared it would only be a matter of time before she died, but we cleaned the wounds, managed to get some liquids into her and a little bit of food mashed up with the water.

It became obvious that she was a fighter and, over time, she began to recover, but it would be a few months until she was back to her normal self, spreading her wings and making an unholy racket. This incident prompted us to get the gander whom we named

Boo (as in 'wouldn't say boo to a goose') due to his initial shyness. Gradually after a few months Boo grew into the gander we always knew he could be and anyone now visiting the farm has our security goose to contend with.

We have had a number of ovine patients including Cyril our ram, who had a suspected thorn in his jaw. It turned out to be a lump and, as it wasn't preventing him from eating or drinking, after a few days he was released. We also took care of the ewe who had twins: because the grass was so bad after the cold winter she wasn't putting on condition and the two lambs were smaller than the others so we supplemented her diet, giving her extra rations. Whiteface Dartmoor are a very good breed for foraging and grazing, but if all the pasture is on one type of soil, it will be lacking in certain minerals. Therefore their diet will need to be supplemented.

As a livestock farmer it is your job to care for your animals and make sure that they have sufficient clean water, enough balanced food and that they are protected from predators. However, a phrase that crops up in farming is that 'if you have livestock, you have deadstock'. One of our ewes gave birth to twins and we kept them in our hospital stall because the male twin had a condition known as pigmouth. This is a common genetic defect with Whiteface Dartmoors, where the lamb is born with a lower receded jaw, giving it a large overbite and making it very hard for the lamb to feed itself. In modern farming you have to think about the amount of time, energy and money you will expend on looking after such an animal and whether it is worth it. On the Edwardian Farm we never leave an animal behind.

For several days we milked Pigmouth's mother, cradled him in our arms and syringed the milk into his mouth and he would make sweet little cooing sounds as he fed. Pigmouth was another animal grateful for the gift of life, propelling himself on his back under his mother in continued attempts to feed. Sadly he didn't make it.

ALEX'S DIARY

I'm a great believer in running a cockerel with a flock of chickens. I find he helps to galvanize the flock, is always on the lookout for predators – a hen-bird is normally concentrating on food and food alone – and provides a constant source of amusement as he fusses over his ladies. The problem I had was that last year's generation of young cockerels was reaching maturity and were starting to challenge the authority of Sunny-Boy, my prize Light Sussex cockerel. I had intended, of course, to eat most of these cock-birds, keeping back only the best bird as a possible replacement for Sunny. Now the time had come for the proverbial 'chop'.

Things came to a head when one hen had been particularly badly harassed and needed to be separated. Sunny-Boy, in his efforts to keep the other cockerels at bay, had clean broken off one of his spurs and was losing blood rapidly. I had to take swift action and washed and bandaged the wound. The other cockerels were contained in a mobile shed and by the end of the week had been sold 'for-the-pot'. We ate one ourselves.

ABOVE My prize cockerel had clearly been up to his usual fighting tricks and lost a spur in the process. Here I am checking the wound for infection.

OPPOSITE Peter and a new born Whiteface Dartmoor lamb as it takes its first look at the world.

THE DARTMOOR PONY

Much of my focus during the Edwardian Farm project has been on our Whiteface Dartmoor sheep and our Red Ruby North Devon cattle, but as a kid the one animal that I associated with the area often referred to as 'the last wilderness of the south' (no – not Streatham) are Dartmoor ponies. These creatures that roam the moor are as beautiful as they are wild and it was my privilege and pleasure to be able to get up close to one of them and join him on his journey from wild moor pony to working farm pony. His name is Laddie and he enriched my life.

Throughout history Dartmoor ponies have been working animals and even in the Edwardian period they were being used as pack horses. However, with the emergence of technology and ultimately petrol-driven engines, their days of working on the farms and delivering goods (milk, mail, coal, etc) came to an end.

For a time there was a market for the ponies in Europe as meat; this is no longer an option due to an export ban on live animals (and the void left has been filled by ponies from Eastern Europe). The temperament of the Dartmoor pony, however, makes it an excellent animal for children.

During the winter the Dartmoor ponies shelter within the valleys; in summer they head to higher ground to escape the flies. They generally have foals between May and August. Their metabolism is designed to cope with Dartmoor, which means they can thrive on quite thin pasture. However, they can graze on the abundant gorse – also known as furze or whin – on the moor (gorse is a legume like clover or beans, and has properties that mean it could be championed as a fodder crop in the modern age).

Gorse flowers all year round but it is best to wait until a plant is not in flower to harvest it. (By the way, next time you walk past the yellow flowering prickly gorse bush, have a sniff of those flowers – they smell of coconut.) To prepare gorse as a fodder crop it needs to be mashed up, to break the prickles and make it more digestible. As a 'flesh-former' in dairy cattle, it leaves root crops such as swedes, turnips, cabbages, mangels and carrots far in its wake; and as a fat-former and heat-producer, it is just behind carrots and still ahead of the other roots.

Gorse's nemesis could be seen as bracken. Bracken is poisonous to ponies if eaten in any great quantity. It is also very hard to control. When Alex and I were working on *Tales from the Green Valley*, we were clearing a field to plant a pea crop. This field was absolutely covered in bracken and initially we began by pulling out the plants by their stems. This merely encouraged growth (as, I believe, burning does on the moor) so, being archaeologists, we took to our mattocks, spades and shovels and spent a gruelling time hacking out the root systems and burning them to

OPPOSITE Laddie, our bright-eyed Dartmoor pony, very quickly gained confidence and it wasn't long before we'd trained him to carry panniers across his back.

CLASSIFICATION OF THE DARTMOOR PONY

Classification and registration of the Dartmoor pony began in 1898 when the ponies were entered in a stud book held by the National Pony Society – formed in 1893 and often referred to as the Polo Pony Society. In 1925 the Dartmoor Pony Society was formed and its primary function is to maintain the stud book. Dartmoor ponies have to be 12.2 hands (127cm) or under and they should be coloured bay, brown, black, grey, chestnut or roan. Piebalds and skewbalds are not allowed (these are similar classifications and both have patchy colour, one of which is colourless or white).

fertilize the field. We thought we had obliterated the plant as the roots we were digging out were like a carpet designed by the Gorgon Medusa. However, bracken seems to also have a secondary root system that is lower than the one just below the surface.

Then one day an old boy wandered through the field and, seeing what we were trying to do, told us the secret to getting rid of bracken is to crush it. You take a piece of wood (like a piece of 2 x 2in/5 x 5cm) with a rope through each end. Hold the ropes in your hand and with one foot on the wood, you progress through the field of bracken. This method bends and crushes the bracken stems, causing a build up of poisons that eventually kill the root system.

The Drift

Every pony on Dartmoor has an owner and animals are marked by one or more of the following: branding, cuts in the ear, tags in the ear or tail design. Each year in late September/early October the ponies are brought in off the moor in a process known as a drift. Bringing in the ponies enables newborns to be marked, the old, sick and infirm to be separated out, some animals to be sold and the rest to be generally checked over.

We were lucky enough to help Charlotte Faulkner bring in her ponies: she wanted to separate out one of her pregnant mares that is known to be a bad mother. This also gave us the opportunity to select a pony that we could take with us back to the farm to break so that we could use it as a working animal – one of our farm ponies, Twinkle, was quite old and not as able as she was. Charlotte's daughters rode off and drove the ponies towards the farm while Alex and I and various helpers formed a line sloping from the high point of the moor towards the road to the farm. We also whooped and flapped when the ponies came into view, to encourage them on their way.

When we had the herd in the yard we isolated a colt from the pack that would make a good candidate for working on a farm. He was three years old, bright eyed, with good shoulders that were laid back and sloping; he was calm, alert and held himself in a good manner with his full flowing tail well set up. His name had to begin with an 'L' to denote the year he was born and we decided to call him Dartmoor Lad or Laddie, as he became known.

Breaking

Back at the farm we had constructed a ring using split chestnut rails, interlocked in an octagonal formation, that was between 20–30 feet (6–9m) in diameter. We could only hope that it was to Mike Branch's satisfaction. Mike Branch is an American (no, scrap that – *the* American) horseman.

A LONG HISTORY

Archaeological evidence suggests that ponies have been on the moor for at least 3,500 years and the first known documented source of the Dartmoor pony is a will of Bishop Aelfwold of Crediton from AD1012. From the 12th century to the 15th century the ponies were used to carry tin from the mines on Dartmoor to the Stannery towns such as Tavistock where it was assayed, smelted and coined. During the Industrial revolutions the ponies' employment opportunities increased although Shetlands were introduced during this period in order to try and create the ultimate pit pony.

He breaks horses for a living – he just happened to be in the area when we needed a chap who knew what he was doing. When Laddie entered that ring he was wild (albeit with a good temperament). By reading Laddie's body language and communicating through his own, Mike was able to get close to the pony, get a halter on him and have him moving as he wished. Finally he managed to back him.

There are elements of mutual trust and dominance, of not presenting yourself front on, of being the first to walk away and of being in tune with the equine mindset.

Alex and I have since carried on his work, spending on average two hours every day in the ring with Laddie. We have had him in the long reins walking on, coming by (left), coming around (right), whoaing, backing and standing still. There have been tense moments but through it all Laddie has been an affectionate and inquisitive pony, keen to learn and sure to be a good friend.

Laddie grazing in his paddock – he proved to be a valuable member of the team.

WORKING A PAIR OF SHIRE HORSES

On my arrival at the farm at Morwellham, I was overjoyed to see that it was equipped with two magnificent shire horses. During my adventures on our Victorian farm I had developed a passion for heavy horses. Having worked Clumper – our Victorian farm shire – in a range of carts and implements, I was keen to try more. The next step up was to get to grips with a 'working pair' on larger implements and undertake more complex jobs. I was quick to enquire about Morwellham's shires, Prince and Tom, to find out what kind of work they had done before – this would be critical for my arable farming plans. To my dismay both horses were adept at pulling tourist carriages but had done little in the way of farm work. I was going to have to set about familiarizing both horses with working alongside each other before we could do any serious field-work.

It is always exciting getting to know the characters of horses, as well as letting them know who's boss and that, if they're prepared to trust you, they'll have an enjoyable time. One thing I have learnt in the short time I have spent working heavy horses is that they actually like work. This is undoubtedly a characteristic that has been bred into them from an early age. With all of the horses I've worked alongside, one of their most admirable qualities is a willingness to get on and throw their shoulders into pulling a heavy weight. I'm sure that in the same way we all enjoy exercise to a greater or lesser degree, Shire horses relish a good work out.

Our first task was to harness up Tom and Prince side-by-side and set about long reining them around the field in which we were planning to grow our crops. For this first training session, they wouldn't be pulling anything but they were harnessed together and a rein attached to the left of the bit in the mouth of the horse on the

BELOW LEFT Megan Elliott, the farm's wagoner, proved a great help throughout the year. BELOW RIGHT Working Prince and Tom as a pair for the first time.

left-hand side – in this case Prince. A second rein was attached to the right of the bit in Tom's mouth. It was a case of getting the basics right, such as getting them to walk forward, to go steady, to go left, to go right, and to stop. They were familiar with all of these commands when working alone pulling a carriage, but the question was whether they would be happy following them when harnessed alongside one another. This first exercise passed without incident and, after a bit of confusion familiarizing them with our voices, both horses went well together.

Chain Harrowing

Feeling confident, we decided to move on to some chain harrowing. A chain harrow is a large and heavy frame of interconnected chains that the horses drag across a field to either 'work down' ploughed soil into a tilth or to rake earth over freshly made seed drills. Permanent pasture can also benefit from chain harrowing, which can help to iron out any bumps and depressions caused by molehills or the churning of the surface by animals in wet weather. We decided to use it to rake out the various small piles of sheep and horse manure that had accumulated in the field so that it could be more evenly ploughed in. Once we started, we soon learnt something of the good and bad qualities of each horse.

To begin with, it didn't take long for Tom to get spooked. After wrapping his nose through the straps that connected the heads of the horses, he very quickly felt uncomfortably constrained and made these feelings known by doing a little dance and attempting to set off. He didn't get far at all, in part because we managed to keep hold of him with the reins, but also because he was tethered to another horse. Prince remained calm and collected and we had an early insight into how things were going to pan out.

As we set about chain harrowing, it was clear that Tom was by far and away the stronger of the two. He was clearly an anxious fellow but he was an obedient and extremely powerful horse. It may well be that as a young horse he had a bad experience with chains – not the chains in the harrow but the draught chains used to pull most agricultural implements. They run from a crook on the collar of the horse, along the side of their body and beyond to the implement they are pulling behind. On a couple of occasions he would find one of the draught chains running between his legs – a rare occurrence when making tight turns. This would cause him to panic, whereas with

ALEX'S DIARY

After a while Prince's casual attitude became exasperating. In one solo schooling session I was close to giving up on him when I decided to raise my voice a little and let him know, in no uncertain terms, that some effort was required of him. Prince was a horse who had clearly had a leisurely upbringing and, while he wasn't a chap to get spooked easily, even he started to pin his ears back and take note, as I raised my voice to a shout. But, all in all, Prince turned into my 'go-to' horse.

In so many ways, it was more about me getting used to his pace rather than him getting used to mine. In fact, some of the more tricky operations in the field required a slower horse and certainly, when I was 'scuffling' (weeding) and 'banking-up' (ridging) in among the potato plants, I needed to be certain that the pace was sure and steady, and Prince was the perfect horse for adding that consistency to the work.

most horses you could simply say 'Whoa', disconnect the chains, straighten up and reconnect. We used to joke that it was a bit like a Formula 1 racing driver being frightened of noisy engines but, seriously, this could be dangerous out in the field and would need some working on.

Prince, on the other hand, was a cool customer. In fact, he was so cool that in the beginning it was difficult to get him to work. When the pair worked together, it became obvious that Tom was doing the greater share and this was most clearly illustrated by the fact that Prince would weave from left to right in an attempt to lighten the load. While heavy horses love to work, they are big and bolshy. If you are not firm with them, they will work out very quickly that if they don't want to do something and they can get away with it, they will.

After a while we got used to the foibles of each horse and as both their and our confidence grew, we all started to get on really well. Prince realized that if he matched

Jack was the farm's run-around horse. Being a Clydesdale, he was a little more lively than the shires and ideal for making short and quick deliveries. Here he is pulling a stock cart used for moving individual animals.

Tom's pulling power and developed a steady pace, he would enjoy his work. Tom became a lot more at ease with his draught chains and the clatter of the harrows behind him. We established a routine of regular sessions throughout the first few months and this expenditure on time would ultimately pay dividends. By the time we'd reached a point where we were happy with them as a working pair, we had probably one of the most chain-harrowed fields in the known world! But this had been invaluable practice for the boys, as their next job was one of the most delicate and difficult in arable farming – ploughing.

Local farrier Ian Mortimer trims Tom's hooves. They needed constant maintenance to keep them in tip-top working condition.

SETTING UP A POULTRY BUSINESS

Towards the end of the Victorian period and into the Edwardian age, poultry and egg production was increasingly being recognized as a viable source of income. William Cook founded his poultry industry in Orpington, Kent and his publication The *Practical Poultry Breeder and Feeder Or How to Make Poultry Pay* was first published in 1882, ran through 15 editions and had sold in excess of 200,000 copies by the middle years of Edward's reign. The book became my poultry bible. It advocated some of the more intensive methods while at the same time respecting the health and well-being of the birds – free-range was very much the order of the day. Cook's methods represent, in my eyes at least, the high point in sound poultry keeping, before the introduction of the appalling practices of battery farming in the post-war years.

ABOVE Regular inspection of the birds to check for lice and condition was important if the flock of laying birds was to stay at maximum production levels. BELOW Duck chicken – a favourite of the farmyard, the chicken that thought she was a duck.

The first step was to select a breed. I'm a Sussex boy and the Light Sussex is a proven winner for both eggs and meat. If I'd been looking at egg production alone, I probably would have gone for a White Wyandotte or a Leghorn. I already had my own breeding flock – one that would put me in good stead for both meat and egg production throughout the year. Although it would have been logical to keep my poultry at the farm, I came across a photo taken around 1905 at Morwellham quayside and behind the cottages I spotted a mobile poultry house. I was excited by the prospect of recreating this Edwardian poultry business and set up two houses. One of these was an old one we managed to wheel down to the location and the second was one that we completely rebuilt from scratch.

PARASITES

One of the main problems with intensive livestock keeping is the increased likelihood of disease and infestation. Conditions in my chicken houses were anything but cramped, the birds roamed freely in the day, but a number of birds roosting in close proximity can cause problems. Then as now, one of the most dangerous risks to a poultry concern is an outbreak of parasites. Regular inspection is crucial and a good poultry keeper is always on the look out for the earliest signs. Lice, mites and fleas can attack brood hens and chicks and their health will deteriorate quickly. In chicks, parasites can stunt growth and make them vulnerable to other problems. They can cause inflammation of the skin through incessant biting – leaving birds looking dejected and emaciated and, ultimately, leading to death.

Specific mites target different areas. Some attack the legs and feet of the birds, burrowing under the scales. Others burrow into the roots of the quills, causing the bird to pluck out its feathers. Lice generally live their entire lives on one host bird – although cross infection should never be ruled out. Mites and lice lay their eggs on the host birds, choosing the soft downy feathers around the rump or under the wings. Fleas jump from bird to bird and quickly spread throughout the flock.

Prevention

Fleas lay their eggs in the dust and dirt in the timberwork of the chicken huts, while some mites hide in crevices during the day then come out to feed on roosting birds. Runs, roosts, perches, walls and ceilings need to be thoroughly cleaned at least twice per year. Traditionally, a mixture of lime dust and soft soap was used: ¼lb (110g) of soap (dissolved in a pint of water/600ml) added to each gallon (4.5 litres) of lime wash. Edwardian poultry keepers had all manner of patented commercial disinfectants and insecticides; some period manuals recommend fumigation using bisulphide of carbon.

For me, the good old-fashioned lime wash represented far and away the cheapest and most eco-friendly method. Regularly changing the litter used in the roosting huts helps. Bracken and straw were plentiful, so I alternated between the two. It is risky to use hay or finer dried plants as these can introduce mites; pinewood sawdust, on the other hand, is said to be so pungent that lice, mites and fleas cannot tolerate it.

The best method for keeping the number of parasites down is to provide the chickens with a dust bath. They often make their own, but I realized that during prolonged wet periods, my chickens had nowhere to bathe. So I set about making them an all-weather dust bath – it isn't as easy as it sounds. I had to dig out channels uphill to stop surface water running into it. I made a roof from an old piece of corrugated tin nailed to six short posts. I dug out the earth underneath the roof and sieved out the stones. The secret is to get the dust as fine as possible. Insecticides can be added to the bath and some manuals even recommended adding paraffin – but I wanted to rely on the bathing skills of my chickens to do the job for me. At first, I was worried they wouldn't take to it but after a few further attempts at ridding it of stones, I noticed the telltale 'bowl' of a used bath and from then on, I would catch them bathing on a regular basis.

POULTRY AND EGG PRODUCTION

When embarking on this project, I was on the look out for enterprises of the day that would bring revenue into the farm. Poultry keeping was one area of the rural economy that was faring rather better than many of the traditional forms of income. In a 1908 assessment of the total value of the output of British farming, livestock, crops and dairy produce accounted for the overwhelming bulk of income for farmers, at more than 90 per cent. Poultry and egg production were valued at 3.3 per cent – more than fruit and flowers (3.1 per cent), timber (0.4 per cent) and wool (1.7 per cent).

The key to financial success in the Edwardian egg business was winter egg production. At this time of year egg-laying can drop off and the dearth in the market causes a rise in prices. Any poultry keeper who could get his hens to lay through winter could stand to make a small fortune. In a more long-running enterprise the trick was to hatch chicks early enough in the year so that they start laying in autumn – while it's still reasonably warm. A pullet (young hen) is the most vigorous layer. The theory goes that if they start laying then and there is enough food around to keep them happy, they'll keep laying all the way through to spring and beyond.

I couldn't test this theory and instead had to find some way of persuading my two-year-old seasoned campaigners to lay throughout winter. I sought advice from both William Cook and the Board of Agriculture and Fisheries leaflets and decided on a two-pronged attack. First I needed to protect my flock from the winter weather and while my houses were draughtproof and watertight, the quayside was exposed to winds whipping up the River Tamar. I set up some wattle hurdles close to the houses so that they both acted as a windbreak and as a sun trap on better days. To encourage the birds to lay I found a recipe from the Board of Agriculture and Fisheries that involved a mixture of raw meat (finely minced beef offal), milled oats and bran which could be worked up, with a warm water, into a mash. To this I added a handful of ground oyster shells and sand to provide both calcium for the egg shells and grit for the digestion. The birds gobbled the mash down and it wasn't long before my small flock of hens provided me with a daily steady half dozen eggs.

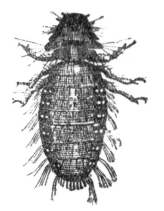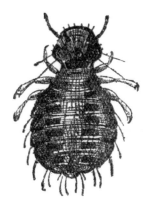

Various lice, mites and fleas proved a constant menace to my flock of chickens.

Breeding Time

With the first signs of spring in the air, it was time to think about expanding my flock. I've always found breeding time stressful: the hens bicker and quarrel for space in the nest boxes, eggs are turned out of nests, chicks get abandoned by mothers and all manner of problems arise on a daily basis. This year I decided to take some tips from William Cook and his book *Practical Poultry Breeder and Feeder*. For the first time ever, I was going to experiment with artificial methods of hatching and rearing.

Incubators had really taken off among Edwardian commercial poultry keepers. I had managed to find an old paraffin-fired incubator but first I had to set aside enough eggs – around 30 – to incubate all at the same time. A paraffin-fired incubator is a fairly simple affair. A lamp sits on the top of a fuel tank and the heat from the flame passes up a flue and into a central chamber in the main box. Heat is regulated through an expanding internal bar which, when it gets too hot, levers a lid on the flue to dampen the flame. The whole affair needed regular topping up with water to keep the appropriate levels of humidity.

I laid each egg delicately in the incubation tray and marked the top of the egg so that I knew when turning them which way up they needed to go. Temperature is obviously key to the successful development and growth of the embryo in the shell – ideally no higher than 103°F (39°C) and no lower than 93°F (34°C). The room temperature was also an important factor, because William Cook advised taking the eggs out and cooling them on a daily basis. This is where I struggled. The danger is in letting the eggs get too cold and killing off the developing embryos. A couple of tips: don't attempt cooling early in the morning – the discrepancy between temperatures

BELOW (From left to right) An external paraffin lamp provided the heat source for my Edwardian incubator; the heat was regulated as it passed into the main chest of the incubator; at all times the internal chamber had to be kept moist and humid.

A relatively small incubator, mine could take 50 eggs at a time.

inside and outside the incubator is too great; the best test is to hold the cooling eggs to your lips – when they feel cool, replace them in the incubator at once.

At seven days it is advisable to check the eggs for fertility by candling – holding the eggs up to a light or candle. If fertile, the shadow of a developing embryo can be seen.

Patience, something I don't have in abundance, is the main requirement of the poultry breeder. The incessant checking of the trays and fiddling with the incubator's settings is really bound to do more harm than good. Incubation should take 21 days and my seventh day check for fertility had already left me with less than half the eggs I'd started with.

I am not afraid to admit that, at my first attempt only a single egg hatched and I was lumbered with a single motherless chick. Fortunately, I was able to place the chick under one of my sitting hens and she duly took care of the little fellow as if it was her own.

Eventually, I had more success and raised 16 chicks from 30 eggs. They hatched all within a few days of each other and were easily transferred from incubator to fostering pen. Once again, I called upon William Cook for advice. During the early 20th century, paraffin 'hovers' were a popular choice. A shallow pyramid-shaped tin roof sits over a paraffin lamp, deflecting the heat from the lamp down on to the chicks. A small curtain hangs from the eaves of the tin roof, allowing the chicks to roam between the hover and a small chick pen. The birds stayed under the hover for warmth until they had fully fledged and were able to cope with cold nights. It wasn't long before they could be introduced to a coop of their own and another generation of chickens began life on the farm.

These four chicks are less than two day old and would require Alex's daily attention for at least the next six weeks.

PRESERVING EGGS

With Alex's poultry concern proving so successful we soon had a glut of eggs. An Edwardian farmer would have needed to think carefully how he sold his eggs for the greatest profit. Sometimes it made more sense to preserve them and sell them later when prices had risen. At times when egg production was a little slow, it could also be a good idea to save them up and sell in larger batches.

The secret to preserving eggs is to keep them airtight. Eggshells are porous, so a way has to be found to seal them, preventing bacteria from growing. The simplest method, and one recommended in the government advice leaflets, is to keep them in a solution of lime water. We had plenty of lime on the farm since Alex and Peter had burnt a large batch earlier in the year.

I put about 2 lb (1 kg) of slaked lime in the bucket and topped it up with around five gallons (23 litres) of water. It stood for ten days, getting a good vigorous stir every morning to ensure that the chemical (and heat generating) process of slaking were well and truly finished. The eggs were then simply wiped clean and popped into the bucket. More eggs could be added at any time.

In this manner eggs can be kept in an edible state for several months if necessary.

Despite its size, the lime bucket was soon filling up with Alex's eggs.

2

ONE FOOT ON THE LAND, ONE FOOT IN THE SEA

Shoreline Resources

 For those who lived at the water's edge there were other resources than shellfish to be gathered. The rocky shores of the south west, and Cornwall in particular, have a colourful history of smuggling and wrecking. By the turn of the 20th century this was long in the past. But disaster at sea still happened, of course, as it does today, and storms could bring their own harvest.

SEAWEED

Seaweed is an excellent fertilizer containing high levels of nitrogen and potash, and is still widely used in gardening. Straight off the beach it also makes an excellent mulch, free from weed seeds, although rather smelly. It can also be dried in great heaps and burnt, the ash being both very concentrated and easy to transport. As ash or as a mulch, seaweed is especially useful on intensively market-gardened land. The high nutrient level replenishes the soil and the complete absence of weed seeds is an important consideration when the ground needs to be kept scrupulously clean by hand. Many people gathered seaweed for manuring the fields. Edwardian coastal entrepeneurs with a cart and horse at their disposal undertook to supply farmers further inland.

ROCK SAMPHIRE

Rock samphire (*Crithmum maritimum*) flourishes along many sections of cliff along the coast. Freshly blanched or lightly pickled, it is a pleasantly crisp vegetable. It is not the same as the more popular marsh samphire (*Salicornia europea*) that was sold for a good price at London's Billingsgate fish market and came mostly from the mud flats of the east coast – as it still does today. Marsh samphire in some ways resembles asparagus in that the very tips are soft but lower down you suck the flesh from the stringy stems. Nowadays you can buy it in season, in upmarket fishmongers. Rock samphire has never had such a large and prestigious market as marsh samphire, but there were pockets of popularity in both Devon and Cornwall where it was sold locally – and people still forage for it today.

The Shamrock, a Tamar barge, moved all sorts of goods up and down the river. Dock dung and seaweed to fertilize the market gardens were but one of her many cargoes.

GATHERING SHELLFISH

The estuaries of the rivers Exe, Teign and Dart in Devon are home to wide mud flats and innumerable wading birds. If you visit them at low tide you will also find, at almost any time of year or day, people scattered along the mud flats. At two or three hundred yard intervals will be a lone figure, usually bent double, gathering shellfish. Except for the Tamar. The mud flats are there and so are some of the birds, but not the people. It was just the same during the Edwardian period. While other estuaries provided food and a reasonable living for the shellfish collectors, the Tamar was already long devoid of such life.

The main culprit was our own Morwellham industry – in particular the copper mine. With the copper came arsenic, which leached out into the waterways and poisoned pretty much every living thing in the lower Tamar. At Devonport, at the river's mouth, the activities of the Merchant and Royal Navy added to the problems.

Pollution problems are nothing new, indeed the Tamar is cleaner and healthier now than it has been for a couple of hundred years. The Edwardian river Tamar below the weir was busy with people, business and tourism, but what shellfish still survived were not safe to eat.

On the other two estuaries, the mud flats are all carefully divided into traditional fisheries, each covering a certain area and each with a specific set of rights to gather. This is not a free for all, but rather a managed, almost farmed resource. By the end of the 19th century most of the rights had been taken under central government regulation, with the Board of Agriculture and Fisheries issuing licences – both to prevent conflict between people and as an early attempt to reduce overfishing.

Out on the coast it was a different story, the cliffs and rockpools had no such regulation. Winkles could be picked from the rocks by anyone wandering along the foreshore, but you had to work hard at it, and know the rocks very well to make it commercially viable.

An Edwardian gentleman Stephen Reynolds regularly went out with an old chap by the name of Uncle Jake along the cliffs around Seaton and he recorded Jake's words in his book *A Poor Man's House*, his account of life in the area (also see page 106):

'Us can walk down to lobster ledge an' west from there to Tattie Rock. I knows where they master gobbets be, if nobody an't had 'em – an' nobody an't. They don' like this iron-bound shop. They leaves it to Jake. But they wuden't, if they know'd what was here... Jest yu turn over thic stone. Ther's some there. My senses, what gobbets they be! If they ther fuddle-heads what goes nosing about Broken Rocks, on'y know'd...'

The two of them were out all day on the rocks from first light. Reynolds had managed to collect two and half pecks (a peck is 2 gallons or 9 litres) of winkles under

Winkles were regularly gathered in Edwardian times.

Jake's guidance, whilst Jake had gathered four pecks, together with a few undersized lobsters. A haul worth around six or seven shillings between the pair of them.

I, however, had a chance to go shrimping rather than winkle gathering. The native brown shrimp lives in the sand close to the shoreline and requires only the simplest of nets to catch. The wooden bar at the base of the net is pushed along the sand; this sends vibrations through the sand which the shrimp can detect. As the bar draws close to them they jump up and out of the sand to escape and obligingly land straight in the net. You work your net up and down the shoreline following the tide over the shrimp-rich sands, up to your knees in water. A heavy swell would make the work very difficult and soak you all over. As it was, even on a calm day, you are wet to the waist for hours on end. I enjoyed it even so and goodness, those shrimp were delicious!

Full buckets of shrimp more than made up for the soaking wet skirts flapping against my legs on the way home.

The Crafts

MAKING LOBSTER AND CRAB POTS

Wherever there are rocky seabeds and sheltered coves, one of Britain's greatest delicacies can be found. Nowhere on these islands is the coast more hospitable to lobster, crab and crayfish than in the south west of England. Not only are conditions ideally suited to the feeding and breeding habits of these prize shellfish, but the natural harbours were the perfect places for fishing villages to develop.

Peter and I were keen to test our sea legs through the winter months and this type of fishing was immediately appealing. Lobster and crab fishing is usually done from small fishing vessels working the shoreline, never too far from the coast. Winter isn't the ideal time to catch lobster as they tend to search for larger prey only after the ambient temperature of the sea has risen above 50°F (10°C). During the winter lobster feed off tiny plankton, but crabs tend to scavenge the whole year round and Peter and I are quite partial to the odd crab supper.

From medieval times onwards, crabs and lobsters were caught at low tide using hoop nets or 'trunks' – these are still used in some parts of East Anglia. However, by the mid-19th century pots emerged as the best way of trapping, and by the Edwardian period this method had become widespread. Few quaysides today are complete without a stack of lobster pots waiting to be loaded up, taken out and dropped, but the style and materials used for contemporary pots are very different from those used at the beginning of the 20th century. Modern steel and polypropylene pots are often made in the parlour style, with multiple chambers and the capacity to trap a number of creatures. Traditional ink-well or beehive pots were far less durable, being made of willow, and many could be lost in a bad storm. As a consequence, when unfavourable winds made

fishing impossible, fishermen spent much of their time in the back yards of their cottages, busily weaving their own pots. My problem was that many of these fishermen had passed away and taken their skills to the grave.

After hunting far and wide, we came across Nigel Legge, a fisherman artist and traditional willow lobster-pot maker who still fished out of the picturesque village of Cadgwith on the southernmost tip of Cornwall. Nigel had spent his childhood fishing with his father and as a boy had been taught how to make traditional lobster pots. He was only too happy to show me the techniques – even though my first faux pas was to refer to them as baskets.

Perhaps the major difference from basketry is that whereas making a basket involves starting at the base and working up, with a lobster pot you start at the top and work towards the base. The basic premise of the pot is to allow easy access to any

Nigel Legge – fisherman, artist and traditional willow lobster-pot maker – not only made his own pots but he actually fished with them.

unsuspecting crab or lobster through the neck at the top and then to prevent it from escaping. The neck is the starting point and 10 to 12 willow rods are inserted into a circular stand at shoulder height. These rods are then woven round with thinner rods to a height of 7¼–8 inches (18–20cm), at which point thirty or so additional rods are then inserted and bent down to begin the bell-like shape. Three vertical rows of plaited rods secure the horizontals and, at the base, the verticals are folded at right angles to themselves, to make the 'chims' (or corners) and base. The weaving of the chims is particularly important as these bear the brunt of strong currents.

In all honesty, it was going to take more than a quick tutorial for me to get to grips with this fascinating skill. I love willow work and have always harboured ambitions of becoming a cottage basket-maker. But these pots need to do more than just carry goods, they need to work and their design is critical to getting a catch and making a livelihood from the sea. I admire Nigel because, like his father, he not only harvests some of his willow locally and makes his own pots – best of all, he uses some of them to actually fish with. Sitting in his cosy shack, chatting about fishing and working the willow, I realized I was in the company of a true craftsman – someone who actually lives and breathes his craft. I only hoped as I set out with my pot that our choice of fishing location and bait would do his pot justice.

ROPE MAKING

There is a clear antiquity to rope making and little doubt that its origins lie very much in man's harnessing of the natural world. From the domestication of animals, the construction of primitive vessels and huts, to the binding of flint to wood to make primitive tools, rope will undoubtedly have played a large part in mankind's technological evolution. Grasses, brambles, reeds and animal hides can all be used to fashion a type of cord, string or rope but certain plants like the common nettle, flax and hemp are especially good for the purpose; their fibrous stems grow straight and tall and can be made in to strong binders when bunched and twisted.

Hemp and flax were the most popular of the plants used to make rope and these grew in abundance on the clay soils of western Dorset, where Bridport was to become a regional centre for the industry in the south west. The plants were harvested and tied together in bundles, which would in turn be soaked in fresh running water for a few days. Removed from the water, they were dried before being flailed to separate out all the fibres in the stalk. The resultant fibres were then 'hackled' or 'scutched' – a process that involves combing out undesirable pith and stalk. It is a very similar process to carding in the wool industry and helps

ROPE MANUFACTURE

The real story of the growth of rope making is in the age of sail when the British Empire was built round vast fleets of naval vessels and merchant ships, and when Britain's coastline teamed with all manner of fishing boats. All of these very many vessels would command millions of miles of rope per year and an important industry grew up around this, with whole areas of coastal towns dedicated to rope manufacture.

to align all of the fibres in the right direction, ready for spinning. At this point too, the fibres would also be oiled with linseed oil or lanolin to make them hardier and more resistant to wet.

With the fibres oiled and aligned, it was then time to spin a yarn. The spinner wrapped the fibres around his or her waist in a thick but loose belt, feeding out, in a twisting motion, the required thickness. The end that was fed out was attached to the centre of a small wheel, which would be spun by an accomplice – usually a child – as the spinner worked away from the wheel. The strength of the rope relies on these yarns being twisted back on themselves, locking them in such a way that they don't unravel. To do this, a rope jack, top and traveller are required. The rope jack is a hand-cranked wheel that drives four smaller wheels: yarn is attached to each via a small hook. The other ends of these yarns were attached to a single hook on another wheel, driven by a hand crank. This was called the 'traveller' by virtue of the fact that, as the yarns are twisted together, their length shrinks and the traveller is drawn towards the rope jack. The 'top' – a conical-shaped block with four equidistant external grooves – could be inserted at the point where the four yarns diverge to create the twisted rope. Gradually, as both wheels at either end are twisted, the rope-maker makes his way along what is called the 'rope-walk', moving the top as he goes. This was the area within which the rope was made: some rope-walks are known to be over 980 feet (300m) long. It has been said that, in his lifetime, a rope-maker would travel the circumference of the globe many times over, in service on the rope-walk.

CORACLE MAKING

The many rivers of Devon rise deep in the heart of this hilly county and flow speedily from their sources on both Dartmoor and Exmoor out to sea, en route cutting deep gorge-like river valleys, treacherous whirlpools and rock-strewn riverbeds. In my short time on our Edwardian farm I knew that I stood little chance of finding the time to build a boat to get out and explore the river. I spent much of my time staring out at what had increasingly become something of a barrier and a 'no man's land' to us. There was, however, one solution to my problem. It lay not in the boat builder's yard of steamed timber and precision shipwrighting but in the willow-beds that lined the water meadows of our farm. Coracles are essentially floating baskets and have their origins in the west of England, Wales and Ireland – in fact, anywhere where there are fast-flowing rivers and streams through rocky terrain. They are recorded as far back as medieval times, but are just as likely to have prehistoric origins, as the materials are so readily at hand and the

Sean Hellman looks on as I massage the main structural withies before bending them in to shape.

technique so simple. They owe their survival mostly to the salmon and trout fishermen of Wales but in other parts of the country they have served a variety of needs. Not only do they offer folk their own means of ferrying across the river; they've been used for retrieving errant sheep when washing them in the stream; and even for catching rabbits marooned on hedge banks during floods. Like many before me, a coracle was a way with which to make use of the river with minimum outlay on time and materials. I didn't want to risk my life on the river Tamar in a 'first-attempt' coracle of my own devising, so I drafted in Sean Hellman, green wood-worker and expert coracle maker.

Sean was quick to point out the many benefits to making and owning a coracle and he soon had me convinced. Firstly, they are easy to make, light enough to carry and despite their shallow draught they are surprisingly stable. They can also be paddled with one hand, which enables the other to hold a net. In the fast-flowing rapids of the traditional salmon and trout rivers of England and Wales they are highly manoeuvrable. Whereas a canoe might very easily find itself jarred against rocks at right angles to a strong current, a coracle will tend to twist and bounce through the gaps.

There are a variety of ways in which to make this dextrous little craft and the various forms are determined primarily on what materials are to hand. Larger willow poles can be rived down into 'lathes' – thin strips of wood – and used to make an open lattice work. A hide – the skin of a cow or pony – can be plastered in tallow fat or lard to make an impermeable membrane to use as an outer covering. We, however, were going to use much simpler materials to make our craft and our first job was to source a few large 'bolts' (bunches) of willow wands or 'withies'. We were, in effect, going to make a basket from these and as our cover, we were going to stitch on a sheet of calico (a strong cloth) tarred in a waterproof mixture made with 6 lbs (3kg) of pitch and a half-pint (250ml) of linseed oil.

To begin with, Sean constructed a mould. This consisted of four timbers fixed in a squat rectangle, with finger-width holes drilled in a circular arc round the timbers. We selected our strongest withies first, to work on the front and back of the coracle. Inserting the butt-end (the thickest end) of the withy into the hole, we slowly bent it round to insert the other end into a corresponding hole on the other side of the mould. Here was, perhaps, the first and most crucial technique of the day. It was vital that the willow was manipulated

PETER'S DIARY

Alex had assured me that his craftmanship was faultless and Ruth insisted her 'Titanic life jackets' would keep me afloat if the inevitable happened. I pointed out to her that the Titanic was designed as an unsinkable ship so logic dictates that the lifejackets were merely a formality and not necessarily an adequate buoyancy aid. Therefore, it was with some trepidation that I embarked upon my voyage on the mill pond in an attempt to catch my trout fry that had escaped from the hatchery.

by hand for, despite its famed flexibility, attempting to bend the wood too soon could break it. The technique requires working the hands up and down the willow, steadily massaging it into the shape you want it to take. This very quickly became second nature but, with even the slightest lapse in concentration, trying to get the willow to bend too soon, you'd find it either snapping in the hand or kinking and irreparably ruining the withy.

Once these primary withies were all in place, we fastened them at their crossing points with short lengths of sisal cord. We then began the weave that held these primaries together, around the edge of the boat. After a couple of rounds, we measured up a bench seat, drilled two holes at each end and inserted it over the primary withies at roughly the middle of the vessel. For the gunwale – the rim at the top of the coracle – we employed a technique similar to weaving rods in a wattle hurdle. Not only did we leave thin rods of willow through the primary withies, we also plaited them around themselves. Thus, very simply, the frame of our coracle was made and as the sun gave up its last light, Sean set me up for an evening of candlelight sewing with a sheet of tarred calico, a needle and some linen thread.

MAKING FISHING NETS

For several centuries Bridport in Dorset had been the main manufacturing area for fishing nets throughout England and Wales. Both hemp and flax grew extremely well, here providing the raw materials. By the end of the 19th century, however, imported hemp was used instead as it was cheaper. Power looms were introduced into the industry by 1900 and factory production of nets was in the hands of just 15 family firms. But while the machines were extremely quick at producing straight rectangular lengths of net, they couldn't produce the many and varied shapes that fishermen required for different jobs. These had to be assembled by hand, partly from machine-made pieces and partly from scratch. Most fishermen customized their own and every fishing family had to have some basic net-making skills to adapt nets and to repair them. The equipment was simple and cheap, consisting of little more than a sort of shuttle – often called a 'needle' – and a stick called a 'spool' to act as a mesh gauge.

How to Braid a Net

First wind the hempen thread on to the needle. Pass it alternately between the fork on the back of the needle and round its tongue in the centre, winding up and down the length of the needle.

Stretch a length of thread horizontally between two anchor points. This will be the top edge of the finished net. Since you will want the finished net to hang down slack in the water, rather than stretched tight, this top edge must be at least a third as long again as the final net will be.

Work a set of loops or 'bights' on to the line. Tie the thread on the needle on to one end of the stretched line – the left-hand end is easiest if you are right handed, and vice versa. Hold your spool up underneath the line hard against your knot and the line, and pass your needle under it, tying a new knot on to the line the other side of it. Now slip your spool out. You have created a loop hanging from the line.

Continue along the line, using your spool to regulate the size of the loops. This first line of work is sometimes called a 'row of half meshes' or a 'round' or even an 'overing'.

Now you are ready to work back the other way doing exactly the same thing, making each knot at the centre of the loop above.

The third round, back in the original direction, will give you your first fully formed diamond-shaped meshes.

Which knot you use is really up to you. There were many variations but the simple sheet bend is probably the quickest and easiest, as well as being historically the most common. To form a sheet bend with your needle; bring the needle up from the back of the work through the next loop, so that it is now at the front of the work. Pass the needle around behind the same loop, leaving the thread loose, and finally thread the needle back down through the original loop.

Nets can be shaped by increasing or decreasing the number of loops, rather like knitting. To increase, you simply make one mesh as usual and then make a second loop or stitch into the same original loop. To decrease, you work your knot through two loops at once.

Nets need protecting from the elements. The salmon fishermen who worked the Tamar preferred to cure their nets with tar. First the net was carefully wound around a pole and dipped into the hot melted tar for a minute or two. A second pole was then inserted into the coil so that between them two people could wring out all the surplus tar while squeezing the excess through the net, ensuring that every bit was coated. The net had then to be quickly and carefully unwound and strung out to dry before it stuck to itself. This process has to be repeated at least every two years to prevent natural fibre nets perishing.

Coal tar was readily available from the coal gas works at Gunnislake, which probably explains why this was the local method of choice rather than the oak tannin or Indian cutch bark that many sea fishermen favoured.

LIFE ON THE WAVES

 For those with a decent-sized piece of land, the sea offered an occasional additional resource, but there were plenty of others for whom the sea was the main source of their livelihood, backed up by no more than a large potato patch.

REAL FAMILY FISHING (A POOR MAN'S HOUSE)

I have turned to this small book again and again: *A Poor Man's House* is an affectionate account of life in the Widger household of Seaton in south Devon in the mid-Edwardian years. Tony and Annie Widger were fishing folk and shared their home with their four youngest children, Bessie, Mabel, Tommy and Jimmy, and a succession of lodgers. It was one of these lodgers – their most long standing, Stephen Reynolds – who wrote the book in 1908.

Stephen Reynolds was a gentleman. For several years he had been holidaying and sea fishing in Seaton and had got to know one of the fishermen particularly well. When his usual accommodation failed him he accepted an offer to lodge with the Widgers. It was a meeting of two different cultures. The book reads a little like an account of living with the natives in some far-flung corner of the empire: the ordinary rituals of life need to be explained and interpreted for the presumably middle-class audience. This, of course, is exactly what has made the book so useful to me. Stephen Reynolds talks about the games the children play, how the dinner is cooked; the nitty gritty of fishing trips and techniques are explained – as are matters of personal hygiene and the household budget. The book was always intended to be a snapshot of life in a poor Edwardian home.

Fishing is a precarious living and like all Edwardian fishing families the Widger's had a number of strings to their bow. Tony Widger fished 'by drifting for mackerel and herring, hooking mackerel, seining for mackerel, sprats, flatfish, mullet and bass, bottom line fishing for whiting, conger or pout, lobster and crab potting, and prawning'. He also made a living 'by belonging to the Royal Navy Reserve; by boat hiring; by carpet beating and cleaning up.' In summer he took gentlemen out on fishing trips and sight-seeing. He even took work pushing wheelchairs. An allotment 'up on land' provided mainly potatoes. The eldest son George had joined the Navy while the eldest girl was out in service. Granny Pinn supported herself on her lace making and midwifery, while Mrs Annie Widger kept house and looked after a stream of lodgers.

Life was quite comfortable by fisherman's standards, with regular meals and boots on the children's feet. While the main breadwinner was in good health they could manage. But it was a life lived on the edge. 'Yu wear's yourself out wi'it an' never gets much for'arder.' With the ups and downs of the seasons, Tony estimated that his average weekly earnings across the year came out at just 15 shillings a week, although of course it came in unpredictable fits and starts. The family needed the regular additional income that the lodgers brought.

One of the things that struck me most strongly reading the book was the degree of overcrowding that everyone took for granted. Although the Widger's house in Alexandra Square is described as 'the largest in the square', it is very thoroughly inhabited. The ground floor consists of just two rooms and a tiny damp scullery. In addition to the six family members living there and Stephen Reynolds himself, there are other lodgers. Often this is just a single gentleman, but for many months the back bedroom – which is described as almost completely filled by a bed – is occupied by a young couple and their toddler who are granted an hour of 'kitchen time' a day to cook their one hot meal.

The fishing was brutal at times. Unlike the fishermen down the coast in Cornwall, the men of Seaton had no harbour. Instead they fished off the beach, which meant that the boats were small and open. They had to be launched from the beach into the surf and hauled up again at the end of the trip. At high and stormy tides they were manhandled over the sea wall and into the street to save them from being washed away. Out at sea there was no protection at all from the weather.

Mackerel were the main fish taken in south Devon, the great pilchard shoals never coming further north than Cornwall. Drifting could bring the biggest catches, but was no easy matter in open boats.

With no modern 'fish finders' to help them Edwardian nets could be empty for weeks on end.

'But there we are – ourselves, the sea, and the heavenly dawn – the sea heaving up to us, and ourselves heaving higher, up and over the top. It exalts us with it. We hardly need to talk. A straight look in the face, a smile...' A POOR MAN'S HOUSE

'Sundown is the time for shooting nets. Eight to fourteen are carried for mackerel, six to ten for herrings – the scantier the fish, the greater number of nets. At Seacombe they are commonly forty fathoms in length along the headrope which connects them all, and five fathoms deep. Stretching far away from the boat, as it drifts up and down the Channel with the tides, is a line, perhaps a thousand yards long, of cork buoys. From these hang the lanyards which support the headrope, from the headrope hangs perpendicularly the nets themselves... nets shot, the fishermen make fast the road for 'ard.'

The men then waited out the night in their open boats. At the first glimmer of light they began hauling in the nets by hand. Some boats were managed by just one man, but two was more common. There was no winding gear aboard, it was just a matter of hand over hand with the sea water running down their arms. The fish were caught tangled in the nets and picked out one by one. Small catches were taken from the nets while still at sea, but if the catch was large the fishermen just bundled it all up and sorted the fish back on the beach.

Stephen Reynolds regularly joined Tony Widger out in his boat as the second man and wrote both about the hardship and the exhilaration of the life.

 ## CRABBING

It is said that in this part of England people have one foot on the land and one foot in the sea. Certainly Morwellham Quay in its time has played host to the ocean-going tall ships of Queen Victoria's Empire and living here we are reminded on a daily basis of our proximity to the sea by the massive tidal range of the brackish river Tamar. Alex and I had managed to get our hands on the plough but we also dearly wanted to get our hands on the tiller and this opportunity came in the form of two briny adventures.

The first was lobster pot fishing with Bill Wakeman out of Brixham harbour. Brixham is famous the world over for its relationship with the sea and has one of the replicas of the *Golden Hinde* (the other one is moored in London) – the vessel that Drake, born in Tavistock, circumnavigated the globe in. Brixham is to this day a working fishing port and its Victorian harbour defences were improved in the early 20th century, demonstrating the importance of the industry in the Edwardian Period. At the turn of the century an ice house was built, enabling fish that was brought off the boats to be taken to markets further afield and sold as fresh. This was one of the factors in establishing fish and chips as a national dish.

However, Alex and I were after crab, which records show have been fished in this country since at least the 12th century (initially using crooks and hoop nets). Alex had made a lobster pot from willow and acquired seven more, and we had placed them in a cove a little way along the coast from Brixham. Modern pots are similar in design and work on the original principle that the crustacean can crawl in through the funnelled entrance but it can't crawl out again. They are made from toughened plastic, which makes them more durable, but this fabric is alien to the prey's environment, whereas a traditional pot is an organic material and blends in. When the pots are placed in the water they take on the taste and smell of their immediate surroundings and it may take a while for a pot that is moved to another cove to become effective – initially it will stand out like a politician at a pop concert.

For someone who works the land looking to make a little extra income from the sea, lobster and crab fishing seems a good way to go.

The major overhead in terms of time is making and maintaining the pots. The rocky environment that the quarry inhabits, coupled with the ebb and flow of the currents, means that the pots may be subjected to rough conditions and can easily sustain damage. Two of our pots looked like they had been attacked by a sharp heavy object. Once you have the pots, it is simply a case of baiting them, weighting them down and placing them in the sea – with a floating marker attached to them for retrieval.

Baiting up can be as simple as taking one of the crabs that you've caught, cutting it in half and skewering it on to a sharpened stick, then suspending it in the middle of the pot. This makes the tasty treat inaccessible from anywhere around the edge of the pot: to access the bait the lobster or crab has to enter via the tapered funnel. Pots need checking fairly regularly. When hauling a pot up you must make sure the line is over the stern of the boat: if you haul away when the rope is over the side, you risk capsizing the vessel.

When we checked our pots we had one or two crab in most of them (it was the wrong time of year for lobster), which would get us a bit of beer money at best, so it wasn't strange to find ourselves sitting in the pub with Bill and his crew. They said that it is not unusual to catch up to 12 crabs in one pot, but fishermen – who are renowned for exaggeration – may get more elaborate with a brandy or two down them. One thing they did mention was not to say the 'R' word while onboard a vessel. The 'R' word being rabbit (same goes for hare). Bill wasn't sure of the origins of this superstition, although it has been mooted that it relates to rabbits gnawing key ropes on vessels or their resemblance to women (who also in the past have been considered a bad omen upon a boat).

OVERLEAF Bill rows back to the crew after a day crabbing. Although not the most bountiful haul of crabs, Peter and I had a great day with a great crew.

TRAWLING

Our next adventure came in the form of an Edwardian trawl. Trawling is a controversial subject when it comes to sustainable fishing in the modern age. However, this is nothing new. In the reign of Edward III concerns were voiced about bottom trawling in 1376. In 1863 an argument was presented in the defence of bottom trawling, likening it to ploughing the soil and saying that the action stirred up food for the fish. Even then this argument didn't hold water.

In this part of the world at the turn of the century (and today), many people's livelihoods would have counted on the catch coming in from the trawlers. Much like mining, sea fishing was speculative, with the spoils first paying off any overheads and then the remainder of the money being divided up in various chunks (the largest going to the vessel's owner), dependent upon role and seniority. It was not unknown to arrive back home owing money.

ALEX'S DIARY

Growing up next to the seaside, I was no stranger to the odd sailing trip and I was delighted to re-acquaint myself with the 'ropes' – if only for a day – on a trawling boat that was nearly a hundred years old. From the outset, the odds were stacked against us. Firstly, our crew had never trawled before. Secondly, we were on a boat that hadn't been used to trawl with for seventy-odd years. Finally, and perhaps worst of all, Peter and I were complete novices to trawling. It was going to be a 'big-ask' to bring in the kind of haul that would have made fishermen a living back in the Edwardian period. But what we lacked in all-round experience, we more than made up for with a mad sense of enthusiasm. In fact, deep down, I was rather sceptical about the possibility of us catching anything at all, but I kept quiet and threw myself ardently into the work.

Silently, we sailed parallel to the south Cornwall coastline. Peter and I leaned overboard and peered at the net-trailing rope, which gradually faded from view into the deep blue of the sea. We grinned with excitement as we talked through our chances of catching a whole shoal of fish.

ALEX'S DIARY

However, when it came to hauling, our optimism soon faded and as the net emerged from the misty blue below it was clear that it was empty. I turned to the camera to present the list of mitigating factors I'd subconsciously prepared. 'We were complete novices' for example and, 'the channel had been over-fished' and, 'we needed at least a couple of practice runs' and, 'the tide was turning' . . . . . when someone shouted, 'FISH!' I span round to catch the slightest glimpse of our quarry obscured by all the fishing tackle and, as we delicately freed it from the net, we realised we had caught the most beautiful skate. Peter and I were jubilant. Buoyed by this early success we turned the boat into the wind, tacked up channel and prepared the trawler for a second run.

The fishing trawler was improved in Brixham in the 1800s. The ships' ochre and oak-bark stained sheets gave rise to the song *Red Sails in the Sunset*. By the end of the 19th century there were 3,000 trawlers in commission in the UK. However, it was the use of the steam engine to drive the capstan, which hauled in the net, that really increased the potential of fishing trawlers and began to impact on fish stocks.

If you are looking for a sustainable way to fish the answer is to employ Alex and me as crew. We boarded the Keewaydin, which was built in 1913, on its maiden trawl. Alex had made some ropes, which we hoped to use and I was set to work painting the vessel – a job that is never done. We also used lanolin to grease up the dead eyes through which the lanyards would pass. The deadeyes are so called because when you look at them without any ropes in they look like the cold dead eyes of a skull.

When we were ready to leave, it was then time to haul up the sail using the call '2, 6' to which everyone on the rope replied 'heave'. This is a way of getting everyone to pull in unison and it is rumoured to come from gun crews pulling out the cannon to reload them. Gun crews consisted of six men but the cannon they were hauling could weigh three tonnes. Obviously they used pulleys like we had on the sail. The term '2, 6' may be an order for men two through to six to start pulling, or it may derive from the French for 'all [pull] together'. Either way, the origins of this phrase are open to debate.

Once the bowsprit had been extended and our sails (which were very heavy) had been hauled up, we set out into open water from Falmouth (the third deepest natural harbour in the world). Pretty soon it was time to drop the trawl net. The net was held open by a wooden beam attached to two metal shoes. We manhandled this over the side and lowered it into the water, watching it descend into the murky depths of the English Channel. We had done a series of repairs to any holes in the net and we could only hope that they would hold and we might catch something.

The trawl lines running from the boat to the net formed a no-go triangle: if they were to snap or the fastenings were to come loose and someone was in this area, they might find themselves being dragged to the bottom of the sea. During one of the trawls the safety line snapped with such force that it seemed to vaporize before our very eyes. Over the course of two trawls we managed to catch a single skate (as I said, Alex and I equal sustainable fishing). However, it was a great learning experience and a good insight into historical trawling. I certainly experienced the relevance of sea songs, because when we were hauling our nets up Lee began to sing and we all joined in with the chorus and this kept us pulling in time.

It is amazing to think how much the sea has shaped us as a nation. So much of our language and the sayings and terminology that we use comes from our nautical past. As a bona fide landlubber it was a pleasure to take part in this endeavour to the bitter end, with the potential of making money hand over fist – even if the crew didn't like the cut of my jib.

OPPOSITE Alex coils the rope he has made prior to setting sail.

BELOW Harriet Thody inspects our entire catch of the day; the lone skate.

FLY FISHING

Having set up a fish hatchery and successfully hatched trout, it was only fitting that I should have a go at the sport that brought so many high-end tourists to the Tamar Valley. It is also the means (using barbless hooks) with which the ghillies at Endsleigh house would catch potential brood stock in order to get the eggs and milt for their hatchery. In addition, the ghillies were also responsible for showing the Duke's guests the best fishing along the river so had to know a thing or two about the pursuit.

I travelled to the Arundle Arms hotel to go and see Dave Pilkington who started off by showing me how to tie a fly. He started tying his own flies at the age of 12 so by now he is very good at it. The fly needs to mimic the insects that are in the air at the time of fishing. To make the fly he used dubbing from a hare's ears, feathers (I brought along guinea fowl and Buff Orpington chicken feathers from our farm and he had among others partridge and old English pheasant), and silk thread that was a hundred years old.

Once Dave had tied a selection of flies, he placed them in a specially designed container that resembled a tobacco tin which had been segmented into compartments, each with its own lid. We then grabbed his split cane rod, his wicker fish basket with a

spring-loaded metal hatch through which the fish could be passed once caught (the basket could hold multiple fish – I always admire an optimist), a priest (not a minister but, rather, a club with which to administer the last rights to the fish), and we headed to the river.

Lesson one was casting. Firstly I was taught an overhead cast and then we progressed onto a roll cast, which is handy if fishing below overhanging trees. Then we began to fish by casting our fly downstream and letting it drift with the current until the line was at full stretch. We repeated this once or twice before letting out a little more line. Then once we were satisfied that we had covered the stretch of river from where we were standing, we moved further down stream.

On approaching the bank Dave stressed the importance of not startling the fish. Noise isn't so much of a problem so it is fine to talk: the worst offence is vibration, which will travel through the water and alert them to the presence of a predator. So we approached each spot on the bank with a light, carefully placed foot.

Opposite where we were fishing were a herd of cows – a mix of North Devons (or Red Ruby cattle) and the paler South Devons. Apparently it was during the Edwardian period while fishing in Devon that Albert Morris had the inspiration to found Ambrosia (in 1917). However, I was too busy concentrating on my casting and the battle of wits that I was undergoing (and by all accounts losing) to focus on the cows.

When cast, the fly will drift over several fish in the river. Trout are territorial, they all have their spots beneath the water and they will all be watching this fly float past. If you are lucky one of them will bite. My first fish? Well, open this book out, measure between the two opposing corners and you might start to get an idea of what I was wrestling with. It may have been a salmon, it may have been a sea trout – all I know is that it got away. We did get a couple of tiddlers which we returned to the water but we only caught one of note. Still I had enormous fun.

ABOVE Mackerel soon to be dispatched Ruth style. LEFT Peter mid fly-fishing lesson trying in vain to outwit a fish.

SMOKING FISH

Once the catch was landed a whole new set of tasks sprang to the fore. Smoking is a traditional way of preserving fish, but by the Edwardian period it was already in decline, as an increasing percentage of the herring, mackerel and pilchard market was for fresh fish. A port with an ice factory and good railway connections could send the catch direct to the London fish market at Billingsgate.

Some fishing families ran their own small smokeries, but generally the smoking was carried out by people specializing in the trade. Smoked mackerel and kippers (a type of smoked herring) were prepared by splitting the fish open and gutting them. They were laid in salt for a day and then hung in the smoke house, preferably over low smouldering fires of oakwood chips. Red herring were prepared in a slightly different way, being simply gutted and not split. They sat in the salt for five or six days before being rinsed off and hung in the smoke house.

The smoke house could be a very simple affair. Up and down the coast of Cornwall you can find a number of odd pits, reputed to be the remains of old smoking pits. Certainly there were places on the north-east coast of Britain where people still smoked herring in pits on the beach. But by now, in the West Country more formal smoking sheds had taken over. Brick or stone structures were preferable, but many were no more than wooden sheds with tar-paper roofs. Earthen floors were quite sufficient for the small low piles of wood shavings that constituted the fires, dotted around the shed.

A plain wooden garden shed makes an excellent smoke house.

Along the sides of the shed were battens to support a series of thin poles. From these poles hung the fish.

After initial salting the fish were wiped clean and a small sharp stick was pushed through the eyes to hold the fish open as it hung in the smoke. A piece of string tied at each end of the stick formed a small loop for hanging. The fish were hung on the smoking pole at 6in (15cm) intervals so that they did not touch, allowing the smoke to pass freely all around them. The fires had to be tended day and night to keep them replenished, but also to keep an eye open in case they flared into life. The smoke needs to be cool and plentiful. If any of the fires burst into flame the temperature would rise rapidly, spoiling the fish and possibly burning down the shed.

Pilchards

Pilchards or sardines, are the same fish at different stages of its life – the sardine being a younger pilchard. They were a staple of the Edwardian Cornish fishing industry and arrived in vast numbers off the coast in a fairly unpredictable way, sometimes disappearing for years at a time. They do not stay fresh for very long so it was essential that these occasional gluts were processed as quickly as possible. Luckily their very oily nature, which makes them so difficult to store fresh, makes them ideal for smoking and salting. Smoked, salted or canned they were exported right across the globe.

Salting was the commonest method. Huge open sheds on the quay front housed rows of women gutting fish with breathtaking speed. The cleaned fish were first salted either in dry salt upon salting floors or in brine tubs, before being packed into barrels with layers of salt packed tight between the layers of fish. The dry-salted fish were often pressed about halfway through the process. The pressed oil made a useful by-product, largely used locally as lamp oil. And with less oil content the fish were able to take up more of the salt, so improving their keeping qualities. The full barrels were sealed and rolled back to the quayside to be shipped off – most went to Italy.

Canning factories were also springing up on the Cornish coast, taking more and more of the catch. Tinned pilchards quickly graced the shelves of most Edwardian grocers shops. Oily fish does well in tins, retaining much of its flavour and nutrients, and sardines or pilchards proved to be one of the cheapest.

With the majority of the catch being processed and packed on the quay, not all that much made its way inland to the local population, but fishermen's families had their own set of pilchard recipes for the mixed bags of fish the men folk brought home.

PILCHARDS

INGREDIENTS

6 pilchards, filleted

Salt and pepper

1 tsp each cloves and allspice
   if you have them

1 medium onion

A couple of bay leaves

½ pint (300 ml) beer

½ pint (300 ml) vinegar

Or instead of the beer and
   vinegar: 1 pint (600 ml) local
   cider

Drunken pilchards

This was another extremely simple dish that made the most of the occasional glut of pilchards. The beer and vinegar cut through the oily nature of the fish.

METHOD

Lay the filleted fish on the table and sprinkle with salt and spices. Peel and finely chop your onion. Lay a spoonful of the onion on each fillet and roll the fish up like herring rollmops. Pack them into a dish, tucking the bay leaves in between the fish. Mix together the beer and vinegar and pour it over the fish. Bake in a warm oven at 180°C or gas mark 4, for around an hour. These baked pilchards eat very well cold as well as hot. Indeed I prefer them cold eaten with good bread and butter.

Stargazy pie

This is the most famous of the Cornish pilchard dishes. It was a rich and special dish reserved for 23 December – not everyday fare. The legend tells of a terrible storm that raged on and on keeping the fishermen of Mousehole in the harbour day after day. It went on so long that food began to run out in many of the fishermen's homes. Starvation loomed. At last, when all seemed lost and still the storm raged, one fisherman, Tom Bawcock managed to make it out of the harbour. No one believed that he would make it back alive, but on Christmas Eve he returned with a boat bursting with fish and the people of Mousehole were saved. Stargazy pie is eaten in memory and celebration of his bravery.

METHOD

First make your pastry. Sift the flour into a large bowl, cut the lard and butter into small cubes and drop them into the flour. Rub the fat into the flour until it resembles breadcrumbs. Add cold water slowly, mixing it into the fat and flour with a cool knife blade. As soon as the pastry begins to stick, lay aside the knife and bring the pastry together lightly with your hand. Knead it a little before returning it to the bowl to rest. Lay a damp cloth over the bowl and set it in a chilly spot.

Wash, gut and bone your fish but leave their heads and tails on. This is called 'pocketing'. Let them sit in a little cold water for a few minutes.

Peel and chop your onions; chop the parsley and bacon. Put all of the parsley and half of the onion together into a bowl. Add the fennel seeds, salt and pepper to the herbs and onion and mix it all thoroughly together. Chop the hard-boiled eggs.

Drain your fish and pat them dry. Now take the herb and spice mix and stuff the fish. Lay them to one side while you prepare the pastry case.

Take two thirds of the pastry and roll it out on a floured surface. Line a dish of around 10 in (25 cm) in diameter with the pastry. Arrange the stuffed pilchards in the dish. Each fish should have its tail to the centre of the dish with its head protruding out over the pastry edge. Now scatter the remaining onion and all of the chopped bacon and chopped hard-boiled egg over the top and between the fish. Pour the cream over the whole lot.

Roll out the final piece of pastry and lay it over the centre of the pie, but don't cover the heads of the fish.

Bake the pie in a preheated oven at 200°C or gas mark 6, for about 20 minutes, then drop the heat down to 170°C or gas mark 3, for a further half an hour.

INGREDIENTS

For the pastry:

1 lb (450 g) plain flour

4 oz (110 g) lard

4 oz (110 g) butter

Cold water to mix

For the filling:

A dozen pilchards

3 medium onions

A large bunch of parsley

8 oz (225 g) streaky bacon

1 tsp fennel seeds

Salt and pepper

6 hard-boiled eggs

1 pint (600 ml) single cream

3

LEISURE AND RELAXATION

Sports and Games

SKITTLES

After our sheep-shearing escapade – or should I say after watching the father and son world-class super-team of George and Andrew Mudge shear sheep faster than I can avoid work – we decided to have a couple of beers and engage in a game of skittles. Skittles is an ancient leisure activity that has been played in Pharonic Egypt, enjoyed by 4th century German monks, written about in medieval manuscripts and depicted in art as a game that was undertaken on the Thames when it froze in winter. However, there is no governing body or generally accepted rules, which has perhaps gone some way to ensure that a number of different variations of the game and are still played in different regions and various pubs.

There are many different names for the game (skittles and nine pins are often considered separate games) and a range of equipment is used. In the past bones have been used as pins. The ball, which is often referred to as a cheese, may be disc shaped as in London skittles or cask shaped as in the Leicestershire game – which allows a greater variation in the movement of the cheese so that more pins can be felled by the skilful player. The alley length alters and the number of throws and how to throw the ball changes significantly. In some games the cheese is rolled the full length of the alley; in others the process of tipping is employed – a player can throw the ball when stood very close to the pins (usually on the final throw). In some games the centre pin must be felled on the first throw and in areas such as the Midlands the ball must bounce prior to striking any of the skittles. Let's not even get started on the 'Dorset Flop' (where the player launches the ball by propelling themselves from a squatting position off a back board along the alley).

The smaller pub game of table skittles is also called 'Devil amongst the Tailors', which refers to the techniques employed by the Dragoons in 1783 when they quelled tailors who were rioting after taking offence at a play that was staged at the Theatre Royal in Haymarket, London.

Although skittle alleys are often found in pubs, the game can be played almost anywhere there is space. We decided to play outside in the apple orchard at Colin Pearce's farm using crudely made ash logs as pins. The West Country, and Devon in particular, is often considered the spiritual home of skittles, and like all other aspects of the game, the skittle pin size and shape changes. The traditional Devon skittle pin is quite large, like the logs that we were using, but this didn't help my aim. Personally I blame poor light conditions, bumpy terrain and the strength of the beverages that Colin was plying us with for my shoddy score (although I did beat Alex).

OPPOSITE AND BELOW
Peter may have beaten Alex in the shearing contest, but he was going to have to knock six out of nine ash logs down if his team was to win at skittles.

Peter marks out the white lines with some much-treasured lime powder.

RIGHT Peter and Alex enjoying a refreshing cup of beef tea.

FOOTBALL

In the early days of football there were no commercial rewards, no gargantuan transfer fees and certainly no £100,000 per week wage deals. The football of the 19th century was an amateur game where fun and exercise were the only incentives for playing and, as such, it was very much a gentlemanly pass-time.

As football became more popular, with games taking place between various clubs and regions, there was a desperate need for a standard set of rules. In 1863, representatives from a dozen southern teams met at the Freemason's Tavern in London's Great Queen Street to discuss a code that could be implemented by all clubs. Thus, the Football Association was born. Eventually, a consensus was reached and on 8 December 1863, the Laws of the Game were drawn up.

Football's official origins owe much to the public schools and universities of the day, whose members made up early major teams such as the Old Etonians, the Wanderers, Oxford University and the Corinthians. Although the Edwardian period heralded a new professionalism in soccer, it was in the first Football Association Challenge Cup tournament that the direction in which the game was heading could first be observed. With the organization and bureaucratic detail taken over by the toffs, it was down to the working-class men of northern England and Scotland to show them how to play it.

Yet, the major controversy that was to develop with the tournament was rather that such competitions in themselves might destroy the very fabric of the game. It was strongly felt that the intense rivalry that formal competitions might create among competing teams would lead to the destruction of the true spirit in which the game should be played in. As early as the 1870s it was clear that success at cup level brought notoriety, fame and, if not a fortune, some tangible financial rewards.

At the dawn of the Edwardian period through to the beginning of the First World War, football was to become almost exclusively a professional game. All of the players representing the national sides of both Scotland and England as well as all of the players playing top-flight football in England were 'professionals', earning a wage from the clubs they

PETER'S DIARY

It had been almost 15 years since I had played a competitive game of football when I found myself setting up a pitch on Dartmoor so that we could face Plymouth Argyle Legends. We marked out the lines, hung the nets, set the corner flags and donned our Edwardian football kits (I blame the decades old boots with bulbous toes and missing studs for my passing inaccuracies, rather than my general lack of skill).

Our manager for the day David Goldblatt complemented Alex and I on our very Edwardian lack of preparation for the game and encouraged us to play rough – employing many tactics that would see the modern player sent off. I tried my hardest to block, hack and shoulder barge the Plymouth players but they just ran rings round me with their far superior talent. The one time I did engage in a bit of shirt pulling, I only did it half heartedly which resulted in a broken finger.

However, it was a fantastic location, a great day and a good break from the farm – and it was made all the sweeter by the taste of victory.

played for. It is also during this period that some of the big names in the domestic games appear. A group of eight teams – Aston Villa, Everton, Liverpool, Newcastle, Sheffield Wednesday, Sunderland and both Manchester sides – dominated the league and cup competitions during the first decade of the 20th century. As prosperity increased among all levels of society, more and more people were choosing to spend Saturday afternoons cheering on their favourite team. Football had made its mark, had become a working-man's game, and was here to stay.

Robust shoulder barging was very much a part of the Edwardian Game.

ALEX'S DIARY

Keen to recreate something of the importance of football to the ordinary Edwardian working man, Peter and I had persuaded a local football club to sign us up for a special one-off fixture. Plymouth Argyle's 'legends' team was keen to play our team – The Ordulph Arms – in an Edwardian-style game. Plymouth Argyle started life in 1886 and by 1903 had turned professional, joining the Southern League. The club grew in stature throughout the Edwardian period and by 1913 were crowned champions of the Southern League.

As the day drew closer Peter and I realised the seriousness of what we'd done. Peter, by his own admission, was no footballer and could barely recall the last time he'd kicked a ball. Having played for a couple of local league clubs I, of course, fancied myself as a canny left-footer with a good passing game and an eye for goal. But it soon became clear after our first training session, that I'd let things slide.

Of the game itself I remember very little. The first pass made in my direction ended up with me simply falling over the ball, and the second pass that came my way went gliding under my foot and out for a throw-in to the opposition. The bulbous toe caps of the

ALEX'S DIARY

Edwardian football boots seemed to impede any form of accurate passing – at least that's my excuse. However, things began to improve and we found ourselves leading 4-2. For me it was time to turn on the style and in what was, quite literally, a last-gasp effort, I whipped in a cross to the far post. Time seemed to stand still as I watched the looping ball float, seemingly unnoticed, beyond the goal, but at the very last second our centre forward leapt like a salmon, rising majestically above the throng of Plymouth defenders, to head home. The victory was complete.

CALISTHENICS FOR LADIES

The average Edwardian was fairly fit by modern standards. Wealthy or not, most people walked a good deal in their daily lives and had some sort of manual labour in their routine – whether that was bringing in buckets of coal, doing the laundry, cutting firewood, or any of the other basic chores. Few people struggled with obesity. Cold homes meant that bodies burnt more calories and there was simply a lot less tempting fatty and sugary food available.

Nonetheless, the Edwardian period saw an increasing concern with physical fitness and healthy exercise. The late Victorian perception of the well-exercised male body as an ideal of hygienic and moral strength began to extend to women as well.

'Calisthenics and gymnastics must not be confounded. We have not as yet quite made up our minds in England whether the latter are an altogether desirable part of a woman's training...'

CASSEL'S HOUSEHOLD GUIDE

Public schools for boys had been at the forefront of a belief in the advantages of exercise for nearly a century. Team sports and cold showers 'built character', and phrases such as 'healthy body, healthy mind' were commonplace in the 19th century. But there had been a great deal of worry about extending these ideas to the female body. Too much jiggling about of the female torso was considered to be dangerous. For centuries medical opinion held that the womb was mobile within the body and that most female complaints arose because it had become misaligned. Although serious medical opinion had moved on, it was still widely thought that jumping off stools (and taking hot baths)

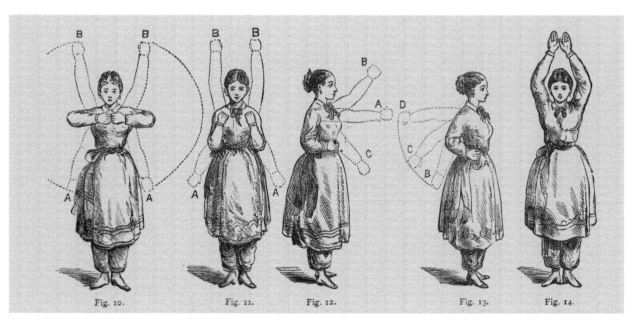

Fig. 10.   Fig. 11.   Fig. 12.   Fig. 13.   Fig. 14.

could cause an abortion, and there remained a general uneasiness, both among the population at large and the medical profession, about athleticism for women.

Calisthenics, the first widely approved exercise regimes considered to be suitable for women and girls, focused on the limbs and shoulders. Long hours of sewing and 'bookwork' led to hunched posture and this was something that it was felt needed correcting. The better posture would, it was hoped, help to open the chest and improve breathing. Cramped living conditions and terrible levels of pollution from coal fires had led to very high rates of respiratory disease. Improved posture and breathing technique was felt to be important in shaking off such illnesses. Calisthenics became widely accepted in schools, and formed a regular part of the Edwardian curriculum, usually under the name of 'drill'.

Although it does look rather ridiculous – and what form of exercise doesn't? – I rather enjoyed the calisthenics. They really did make a significant difference in loosening and relaxing my shoulders after a hard day lumping buckets of water and coal. I also found it helpful to do a few of the exercises as a break from sewing.

This was the beginning of a whole industry. Aerobics, fitness videos, yoga classes and gym membership all owe their popularity in Britain to the brave new world of calisthenics. The fact that you could pursue health, fitness and the body beautiful in private made it all the more popular. This was the Wii Fit for the 1900s.

'The constitution must be coaxed, not strained; the strength not unduly taxed, no over fatigue ensuing; for exhaustion makes people look worn and old.' CASSEL'S HOUSEHOLD GUIDE

Cassel's Household Guide *gives full instructions for Calisthenics in the home.*

Fig. 15.   Fig. 16.   Fig. 17.   Fig. 18.

BICYCLES

Bicycles were a Victorian invention, but it was in the Edwardian era that they really came into their own. This was due to improvements in manufacturing bringing the price down within reach of an ever-growing number of people. But it also had something to do with new attitudes. Early on, bicycles were not considered to be suitable for women. Scare stories circulated about how riding such a machine could make a girl unfit for marriage.

But as the new reign began such fears were fading. Bicycles were much cheaper than a horse and now, with a few modifications to the shape of the frame and the addition of guards on the chain and wheel spokes, a woman could ride a bicycle without any adjustment to her usual dress. Fanny Bloomer's radical garment – named after her – was no longer needed. Although a pair of bloomers was a practical garment, those who had been brave enough to wear them in public had been subjected to public humiliation and even assault. Now the world had moved on.

ABOVE The skirt guard over the chain was invaluable for women cyclists.
BELOW AND OPPOSITE Bicycles were becoming increasingly affordable. Edwardian women took to the bicycle with gusto as it gave them so much freedom. I felt like the world was my oyster when I was given a bike to ride.

A woman on a bicycle was a woman taking healthy outdoor exercise, and household manuals even included instructions on learning to ride. Fancy trying to learn from the following, taken from *Consult Me – For All You Want To Know* (1901):

> *Bicycle, to Ride. – The first time you try should be in a boarded room, or on a gently falling smooth road. Mount the machine from an elevation, and slightly pushing off with the inside foot, let the bicycle run without placing the feet upon the treadles. This is a simple way to practise balancing, and the moment the machine appears to be falling to the left, draw the left hand a little towards you, it will have the effect of turning the driving wheel in the same direction, readjusting the balance; if it should fall to the right, reverse the action, and guide the wheel to the right.*

> *When you have mastered the preceding, bring the bicycle up to the side of the road, let the outer treadle be slightly inclined or just falling over the centre, then throw your leg over the saddle, press on it, at the same time pushing off from the side with the inner leg, by the time you have done this the inner treadle will have come up, and as soon as it has begun the descent place your inner foot on it, and press down as you did the outer one. Give the treadles a steady pressure and then only when going over, allowing the wheel to complete the revolution itself, the foot merely feeling the touch.*

Men too saw the opportunities a bicycle could bring. Here was a form of personal transport that those at the upper end of the working classes could afford. In some areas of the country funds were set up to help unemployed men buy a bicycle so that they could travel to new jobs. Once you had a bike there were no further costs in using it. The railways had opened up the country, but train tickets were expensive. A bicycle made regular Sunday outings a real possibility. Picnics at local beauty spots or a tour of the neighbouring village pubs were popular bicycle trips – especially with the young. Few of the older generation learnt to ride, so a group of young people out for a ride was a great opportunity to flirt without the oldies looking on, and many a couple courted by bike.

RUTH'S DIARY

My bicycle was a wonderful machine. Extremely sturdy, but not as heavy as I had expected. The real difficulty I had with it was how high it was. There was no way to adjust the height of the saddle. It was a good six inches higher than a modern bike and this made getting on and off rather tricky. Once on it, however, it was great. Sitting up so high it felt really fast, and downhill it certainly was. With no gears, uphill was serious work, I puffed like a train and advanced at a snail's pace. On the flat it went like a dream, eating up the miles. I quickly learnt to walk up the hills and just enjoy the rest of the ride.

The skirt guards were amazing. Not once did I get tangled up, nor did I find any grease from the chain on my clothes. To my surprise, the brakes were up to modern standards – very efficient and reliable – and the bell was frankly the best I've ever come across.

It was easy to see why Edwardian women took to the bicycle with such gusto. It gave such freedom. Although the railway had got as far as Calstock, that was on the other side of the river. The nearest public transport remained six miles away. With a bicycle suddenly my options were that much greater.

LETTERBOXING

Standing at the foot of a tor with the wind driving the rain into our faces, I watched bewildered as the woman who would become my wife scoured the cracks in the imposing granite rock looking for what she called 'letterboxes'. Personally, I thought she was mad and as the weather was coming in (despite being the height of summer), I thought our time would be better suited getting off the high ground (and into a pub). However, her search was fruitful, and pretty soon I was joining in the hunt.

Once I had found my first letterbox I was hooked on this peculiar pastime that began in the mid-19th century when railways reached Devon, making Dartmoor accessible and sparking a boom in tourism. Due to the moor's inhospitable nature and constant weather changes, the services of a moorland guide were essential. One of the most respected guides was James Perrott, who was born in 1815 in Coombe near Gidleigh Mill and moved soon after to Chagford where he remained until his death at the age of 80.

Alex, Peter, Robert Paul and Rupert Acton out to conquer Cranmere Pool – the original letterbox.

LETTERBOXING AROUND THE WORLD

Letterboxing has now gone global, with a strong following in America, following a newspaper article about the curious hobby that began in south west England. It all started on Dartmoor, so next time you are out walking in this beautiful national park have a look in the nooks and crannies (always remembering never to do any damage) and you may find a letterbox. If you don't, you aren't looking hard enough (trust me, it took me awhile to get my eye in to spot the likely hiding places). And if you visit Cranmere Pool where it all began, you will see a permanent stone box built following an appeal by the Western Morning News in 1937, commemorating James Perrott, the man who started the craze.

OPPOSITE The letterbox at Cranmere Pool; the most isolated place on Dartmoor and site of the first letterbox in 1854.

One route that he took tourists was from Chagford to Cranmere Pool, one of the most inaccessible areas of the moor (see right). It is surrounded by peat bogs, receives on average more than 100 in (254 cm) of rain a year and, due to the undulating landscape, it can be easy to miss the pool – even if you are stood within a few hundred feet of it. The outward journey of around eight to nine miles (13–14.5km) started on horseback and concluded on foot. It was here in 1854 on the banks of Cranmere Pool that James Perrott first placed an earthenware bottle in which he left his calling card so that others could contact him and also leave their cards. And so began letterboxing.

The evolution of letterboxing has been slow but exponential. In 1888 a tin box replaced the bottle at Cranmere Pool and a system developed, where postcards were left and when the next soul came along they would gather the contents and post them, while leaving their own postcard. A second letterbox was established at Taw Marsh in 1894, and a third at Duck's Pool in 1938. In 1905 a log book was included with the Cranmere letterbox so that there would be a record of all visitors to the letterbox sites.

On 22 July 1907 the Edwardian John H. Strother suggested the inclusion of the rubber stamp system still in use today. He wrote:

'Reached the pool at 7:10pm, misty day with cool breeze, and would suggest that a rubber stamp, something like the post office stamps for postmarking letters or rubber stamp for putting the address at the top of a piece of notepaper be provided and kept here. If this were done it would be proof that cards posted had really come from Cranmere.'

Up until 1970 there were only 15 letterboxes on the moor, but then this number began to increase and the pursuit exploded in popularity in the 1980s. Today you might find several letterboxes on a tor, as well as elsewhere on the moor. Each one is a weatherproof container holding a stamp and a log book and usually an inkpad (although you should bring your own to be on the safe side). Above all else, they will be well hidden. The idea is to collect as many stamps as you can and leave a note to the owner of the letterbox identifying yourself, along with the date and an imprint of your own stamp.

Edwardian Inventions

History is not something that can be fully quantified or compartmentalized, especially when dealing with multiple factors and independent-minded inventors. However, an individual event such as the submission of a patent has definite parameters. We may not know the thought processes that went before, the accusations of plagiarism that went after or, indeed, the fortunes of the invention, but we do know that the paperwork was submitted and the named inventor was credited with the creation.

Patents are applied for all the time, but there was definitely a rush on during the Edwardian period, thanks to the availability of both the internal combustion engine and electricity, coupled with the emergence of the middle classes from an upstairs-downstairs culture. For example, during the years between 1901–1905, more than 140,000 British patents were granted (which gives an idea of how many were applied for). This is the era that saw the car and the plane really come into existence and, some may argue, this was the true beginning of the problems we are dealing with at the moment.

During our time on the Edwardian farm, we came into contact with a number of these inventions. On our mining expedition, for example, we used a device similar to the 1904 Simon's Improved Hook to attach our carbide lamp to the face of the rock. Also, I was visited by an Edwardian travelling salesman with a seemingly bottomless suitcase of home-help devices. His name is Maurice Collins and he collects antique gadgets. In his seemingly bottomless suitcase he had a portable moustache protector; a beard net to wear when asleep; an egg sorter (the eggs rolled down a frame and dropped through spring-loaded trap doors, dependent upon weight); a self-pouring teapot (invented in 1886 and popular for a 30 year period, selling in tens of thousands); and a burglar alarm. My favourite was the clockwork 'teasmade' patented in 1902.

It is made out of brass and copper and, once activated, a match was struck, which in turn lighted a burner to heat the water and tea in the pot until boiling. After sufficient time had passed, a catch was released and the platform on which the pot was sat was raised and the tea poured into a waiting cup. Simultaneously the device would extinguish the burner flame and wake you from your slumber to enjoy the first cup of tea of the day on your bedside table. During the Edwardian period it was possible to buy gadgets like this on a pay-monthly system (which obviously means paying more than buying outright); but in a world of varied incomes from enterprises such as market gardening, this was a useful scheme. Also Maurice is quite the salesman.

ELECTRICITY

Throughout history civilizations have been aware of the presence of electricity – be it in the form of electric eels, a static electric shock or a lightning bolt. In the 1600s, when Englishman William Gilbert started to study magnetism and electricity, the phenomena began to be quantified. He distinguished between static electricity and the lodestone (magnetic) effect.

Around the turn of the century mankind really began to harness electricity by building power stations. Initially many were driven by water, such as the one at Niagara Falls, but later they incorporated gas turbines and by 1920 the first power station to be powered purely by pulverized coal had been built. Along with a new source of power came a new range of technology. Devices including household electrical items like the vacuum cleaner and the washing machine turned up on the scene in 1903.

THE EVOLUTION OF ELECTRICITY

The study of electricity progressed steadily throughout history. Otto von Guericke invented a static electricity machine; Robert Boyle realized electricity could be transmitted through a vacuum and observed attraction and repulsion; Stephen Gray discovered conductivity; and Charles Francois du Fay recognized that electricity is either 'resinous or vitreous' – later defined as negative and positive.

Benjamin Franklin probably conducted the most famous experiment in the history of electricity when he tied a key to a kite and flew it in stormy skies in 1752 (he later invented the lightning rod). Orsted, Amperè, Faraday and Ohm all advanced the science in the early 1800s. Later in the century, Maxwell, Tesla, Edison, Brush, Siemens, Bell and Parsons to name a few, developed electricity into something that could advance industry.

FLIGHT

Along with the understanding and harnessing of electricity came the race to escape the confines of the earth's gravitational pull, via powered flight with control through all three planes of direction. Discounting legends such as Icarus and the drawings of Leonardo da Vinci, the first manned ascent was by means of the practical hot-air balloon built by the Montgolfier brothers in France in 1783. They believed they had discovered a new gas but they had merely stumbled upon the fact that air gets lighter as it heats up.

Fig. 44. Franklin et son fils faisant, à Philadelphie,
l'expérience du cerf-volant électrique.

The next in the quest to conquer the skies was Sir George Cayley (1773–1857) who is considered the father of aviation by many. He identified the principle forces that were involved: weight, lift, drag and thrust. In 1857 the Frenchman Félix du Temple (de la Croix) patented a design for an aerial machine and in 1874 he flew a monoplane in Brest down a slope which bystanders affirmed achieved lift off.

In 1890 on 9 October, Clément Ader managed to fly 54 yards (50m) in his craft known as the Eole at a height of around 6in (15cm) but most importantly on flat ground. He continued his work and unconfirmed reports began to emerge that his new machine the Avion III had flown around 328 yards (300m) – this was later refuted. Also in the 1890s the German Otto Lilienthal – 'the glider king', whose work was an inspiration to the Wright Brothers – was the first man to launch, fly and land a craft. One of his longest glides was just shy of 383 yards (350m). Lilienthal was later killed when one of his gliders crashed.

At the beginning of the 20th century flight really began to take off. In 1903 the airship built by the Lebaudy brothers flew 37 miles (60km). Later that year on the morning of 17 December, Orville Wright piloted the Wright Flyer on a 40 yard (36.5m) flight for a period of 12 seconds at Kill Devil Hills in Kitty Hawk, North Carolina. This is recognized as the first powered, fully controllable, sustained and manned flight in a heavier-than-air craft.

The following year ailerons – controllable flaps – were incorporated into a Wright Brothers inspired glider by the Frenchman Esnault-Pelterie and in 1908 Wilbur Wright won the Michelin prize for flying 77 miles (124km).

In the West Country there is talk of the son of Cornish immigrants to New Zealand, Richard Pearse. Claims have been made that the monoplane that he built, which is similar in design to modern aircraft, flew nine months earlier than the Wright Brothers. All we can say for sure is that in Edwardian Britain travel took on a third dimension.

OPPOSITE Benjamin Franklin depicted conducting his kite and key experiment reputably in the June of 1752 – he would later invent the lightening rod. BELOW Peter with the Bleriot aeroplane.

The Tourist Industry

The seaside holiday was fast becoming a fixture of British life. Resorts with good railway links to the great manufacturing cities had seen decades of explosive growth, as more and more workers were able to afford a week by the sea. Day trippers helped swell the numbers. A new seaside culture of food and entertainments had arisen to cater for them. During the season thousands of people a day arrived at places such as Skegness, Margate and Blackpool. From the 1860s to 1880s the growth had been piecemeal, but from then on local government took an increasingly important hand in the development of seaside resorts. Councils began to regulate and inspect establishments to ensure that the customers were not being short changed, hoping to secure their town's reputation and ensure repeat business. Parks were created, the sea front improved with tidy walks, benches, bandstands and so forth. And where private enterprise hadn't endowed a resort with a pier, several town councils stepped in to build them.

By 1900 the seaside resorts of Britain were pretty much well-established. Business was booming.

The Corporation at Torquay in South Devon, for example, laid out and maintained a recreation ground with running and cycle track, and built a pier in 1895 at a cost of £10,500, with a pavilion that hosted concerts, flower shows and, by 1910, roller-skating! The King's Gardens were laid out in the first year of Edward's reign and included a large pond for children to sail their toy boats. The esplanade was planted with subtropical species to link the various amenities. The Princess Gardens adjacent to the new pier had two bowling greens as well as tennis and croquet courts. In all the Torquay Corporation spent close to £200,000 providing visitor facilities – a truly vast sum in an era when a schoolteacher could earn £35 a year.

Devon and Cornwall attracted a distinct type of holidaymaker. Their distance from the main manufacturing centres made them too expensive for most working-class holidaymakers to travel to. Those who came to the south west tended to be fairly wealthy and also stayed longer. Factory workers were restricted to an annual week's leave, while the better-off tourist had more freedom. A surprisingly large number 'took

a house' rather than lodgings and stayed for much of the summer, with the man of the household making a number of business trips back up to London, leaving his family behind at the seaside. Lodgings too were generally larger and grander than those found at most other resorts.

In tourist guides of the time, seaside towns in Devon and Cornwall made a point of emphasizing their genteel credentials, claiming that they were entirely free of those more vulgar amusements such as Black and White Minstrel Shows and Punch and Judy booths, even though photographs included in the very same publications show Punch and Judy were alive and well in both counties.

The seaside attracted by far the largest number of tourists to Devon and Cornwall. Ever since Charles Kingsley had published his novel *Westward Ho!* in 1855 the coast had become a subject of enormous interest to the reading public. But it was by no means the only tourist destination. R.D. Blackmore's *Lorna Doone* (published in 1868) and then later sir Arthur Conan Doyle's Sherlock Holmes story *The Hound of the Baskervilles* had made Dartmoor a major attraction in its own right. Coach trips ran most days from Bovey Tracey, Moretonhampstead, Ashburton, Tavistock and Okehampton. At the beginning of the Edwardian period horse-drawn coaches followed a regular circular route but as the decade moved on the horses gave way to motorized charabancs. The extension of the railway to Princeton brought many more people right up on to the moor.

The Edwardian love of picnics was one which became even more appealing for those who could afford a car. Motoring holidays were hugely popular amongst the wealthy and brought tourist income into even the remotest parts of Dartmoor. We were immensely lucky to be invited to such an occasion.

Whether you had a chauffeur or chose to drive yourself, motor cars transformed holidays for the upper classes.

Enjoying the natural beauty of the countryside was the major aim of visitors and Ward and Lock's illustrated guide books contained times, prices and routes for railway excursions, coach trips and trips by steamer. They also included extensive walking tours with hints on the best roads for cyclists and motorists.

'This is a favourite run with cyclists who can face forty to forty-five miles with equanimity... ask for permission to wheel through the grounds of Trelowarren Park... Thence over Goonhilly Downs, a flat breezy rerun, the road usually perfect... the finest road for returning is undoubtedly the same main Helston coach road... we reach a fork with no direction post. Road on right of fork leads (one mile) to Traboe but the road thence to Manaccan is hilly, difficult to find, pretty, but not in first class order...'

As well as admiring the views visitors were also expected to take an interest in the history and architecture of the area. Notable churches feature heavily.

'The church of St Neotus (1321) has a magnificent series of fourteenth-century stained-glass windows, famous throughout the West of England. These windows, having become dilapidated through neglect, in 1825 the Rev RG Grylls, of Helston, began a thorough restoration. It was nearly five years before the work was finished, and the expense was over £2,000.'

Interest from tourists over the years was increasingly encouraging other individuals and institutions to have regular opening times for visitors. For centuries the gentry had been free to knock on the door of great houses up and down Britain to ask permission to look around. If the family was away they usually left instructions with their housekeepers about whether or not visitors were to be admitted. Over time the middle-class tourist had come to see this almost as a right. In the Edwardian period more and more 'places of interest' were formalizing their opening arrangements and publishing them to an ever wider group of people. Any reasonably tidily dressed visitor could wander about asking questions for the price of a loaf of bread – or even, in some cases, for free.

Hotels were springing up all over to accommodate these visitors and the provision of 'teas' was a useful additional income for many moorland farmers.

Cotehele house on the Tamar was listed in the 1894 *Tourist's Guide to South Devon* as open upon application to the Manor Office, Emma Place, Stonehouse. Application could be in person or writing. By 1908 when the *South Devon and Cornwall Thorough Guide* was published, regular opening hours were listed and the guide further points out that 'the intelligent cicerone who shows the house makes no charge for his services, but well earns a tip.'

It wasn't only country houses and castles that the Edwardian tourist could visit. The Lizard lighthouse was open on weekdays from 9.30am until an hour before sunset, but not before noon on Mondays. 'The officials who conduct visitors round will explain the working of the machinery,' the guide book stated. And of course there were museums. The Royal Institution of Cornwall had its museum in Pydar Street, Truro where visitors were encouraged to view objects from the 'flint age' as well as prehistoric burial remains and a pair of cannon balls fired from Pendennis Castle.

WHAT THE TOURISTS MEANT FOR US

Our quayside saw its own regular flow of tourists in the Edwardian period. Five different steamer companies ran regular trips up the Tamar in summer, carrying up to 2,000 people at a time. 'The grandest part of the Tamar lies between Calstock and the weir head... between Morwellham and the weir head the river flows through a deep gorge,' says *Worth's Tourist's Guide to South Devon*, while the Baddeley and Ward guide mentions: 'On reaching Morwellham Quay the tourist may ascend to the summit of Morwell Rocks.' The local daily coach tour from Tavistock also included Morwell Rocks in its itinerary.

The steam boats moored up at the quay for a couple of hours (the Duke of Bedford insisted on mooring fees as his cut of the trade), after making their way up the river from Plymouth. The stop allowed visitors to take either the gentle walk along the Duke's

PREVIOUS PAGE AND RIGHT Fruit and flowers from the gardens and hedgerows were quickly prepared for the influx of tourists. Children were a very successful sales force.

Drive by the side of the river to the base of the rocks or, for those with more puff, to take the steeper path to the top. Once there people admired the view and tried out their shouting and yodelling skills on the echo.

This was an excellent opportunity for the local population to cash in on the tourist boom. With only one small inn in Morwellham, there was plenty of room for a little private enterprise. Teas were in demand. Local market garden produce was also ideal for selling direct to day trippers – soft fruit, a bunch of flowers, maybe a nice pot of jam to take home.

Tavistock was marketed to tourists as both a centre for visiting Dartmoor and as an ideal location for fishing. Several hotels and lodging houses catered to visitors, who were also keen to eat well while on holiday. The summer demand for top-quality fruit and veg, and for poultry and dairy products was high.

Ewert Hutchings in his autobiography recalls his parent's involvement in this business. On Tuesdays his father hitched up the horse and cart and began a regular round of local farmers collecting up dairy produce, rabbits, eggs and poultry. On Wednesdays he delivered the produce to a regular round of hotels, clubs and restaurants with any remainder being taken to market on the Friday.

Like so many Edwardian rural dwellers the Hutchings survived on a patchwork of enterprises.

OVERLEAF Five different paddle steamer companies ran regular excursions up the Tamar to Morwellham in the Edwardian period, some carrying 2,000 people at a time.

HOTELS

The best hotels in Edwardian Devon and Cornwall were keen to boast of their facilities. They wished to attract the rich and fashionable, so offered the very latest in domestic facilities – a world away from the way most locals lived. The adverts carried claims such as 'bathrooms on every floor (h.&c.)' or offered 'the finest sanitary arrangements'; 'electric light in all rooms' turns up repeatedly and also, though less often, electric or hydraulic lifts to all floors. These were recognizably modern hotels, unlike the more spartan boarding houses for the less wealthy holidaymaker. By 1904 many were also offering motor garages and inspection pits for car-owning customers to store and repair their expensive and somewhat unreliable machines.

The finest of Devon and Cornwall's local produce poured into hotel dining rooms at premium prices.

Sidmouth

SIDMOUTH (South Coast), DEVON.

THE VICTORIA HOTEL

SITUATED near western end of Esplanade, near to Brine Baths, and conveniently to Golf Links. Commands magnificent Sea and Coast Views. Surrounded by fine well-sheltered Pleasure Gardens, Walks, and Terraces. Perfect Sanitary Arrangements. Electric Light throughout. Passenger Lift Specially designed for Invalids. Large and commodious Bedrooms and Private Sitting Rooms; handsomely furnished. Dining and Drawing Rooms; spacious Smoking Lounge, Reading, and Billiard Rooms. Excellent Cuisine and well-selected Wines. Outside Iron Staircase Escapes. Motor Garage, with Inspection Pit. In consideration of subsidies paid by the Victoria Hotel, its guests have special privileges at the Sidmouth Brine Baths and at the Golf Links, 20 per cent. being allowed off their charges or subscription. Nat. Tel. 11. For Terms, Particulars, Illustrated Guide, &c., apply to
JAMES MACGUIRE, MANAGER.

CLOTTED CREAM

Devon was already famous in Edwardian times for its clotted cream. The tourists who flocked to the region in the summer were encouraged to look out for it in the guide books. Stories were told of the king making incognito visits to humble homes to sample the delights of the tea table. It was even possible to have cream posted home for you.

Most tourists enjoyed clotted cream served at their lodgings or hotel, or maybe at one of the small cafes in the more popular resorts. At such establishments clotted cream was usually served with strawberry or raspberry jam and scones, along with a pot of tea.

For the forward-looking Devon dairy woman there were two main routes open in the first few years of the 20th century. One was to increase her clotted cream production and become a regular supplier to as many hotels and other catering establishments as possible. The other was to take government advice and move into soft cheese production. Good-quality small soft cheeses could find a market among the visitors but also in higher-class shops and establishments in London. The main considerations in both cases were neat and attractive presentation and a reliable supply.

With so very few facilities at my disposal I decided to restrict my clotted cream production to that which I could sell to the visitors who came off the river steamers to view Morwell Rocks – and, of course, for our own consumption. With only the most basic of spaces available to use as a dairy, and no good ripening cellar or room, it was important that I didn't bite off more than I could chew when it came to making cheese. The simplest of soft cheeses, the 'Cambridge' style cheese, looked to be the one that would be most successful in the circumstances. It was known to be a fine accompaniment to soft fruit in the summer months, so hopefully would be saleable alongside other Tamar Valley produce.

Clotted cream by post quickly established itself as a tradition, encouraging tourists to enjoy their holiday treat all year round in their own homes.

TORQUAY.

Devonshire's Speciality & Peculiarity.

RICH CLOTTED CREAM

FROM

The Old Tor Abbey Dairy, 5, Lucius Street.

Proprietor—F. BLATCHFORD.

SAMPLE TIN OF CREAM SENT TO ANY ADDRESS ON RECEIPT OF 1/6

FAMILIES WAITED ON.

HAMPERS OF DAIRY PRODUCE

AND

Guaranteed Prime Devon Poultry & Game

SENT TO ALL PARTS.

PRICES ON APPLICATION.   Telephone 402.

4

AN AUTHENTIC EXPERIENCE

The more adventurous tourist who ventured a little off the beaten track could search out a more 'authentic' experience by having tea at one of the more remote and traditional farmhouses. Dartmoor was considered to be an especially good location. Scones were very unlikely to be served in a farmhouse; rather it would be the local 'splits' (see page 150). The cream too would taste different.

Cream absorbs other flavours very easily and the flavour it was most likely to absorb in a farmhouse kitchen was smoke. Up on Dartmoor they mostly burnt peat in an open hearth.

The cream was made on the fire and took on a distinctive peaty flavour. Down in the valleys the older and more picturesque dwellings still cooked on open hearths as well, but here they were burning wood, giving a different flavour entirely to the cream. Such open-hearthed kitchens were very attractive to the well-heeled tourist, who enjoyed the 'olde worlde' feel. They wrote about their 'charm' in letters and postcards home and they tried to photograph them, despite the low light levels.

MAKING A CAMBRIDGE CHEESE

 Each cheese requires 8 pints (4.5 litres) of full-fat milk, although it is best not to make only one at a time. A minimum of four cheeses should form each batch. If the volume of an individual cheese is too small, it loses heat much too quickly and the texture of the curd is watery and unpleasant.

The milk is gently warmed to between 92–95°F (33–35°C). It should lie in a wooden tub of no more than twice the volume of the milk in size, and one with a snugly fitting lid.

By the early years of the century rennet was freely available in a purified and liquid form. It was no longer necessary to prepare a calf's stomach; you simply bought a bottle from the dairy equipment supplier.

One dram (4ml) of rennet could set two gallons (9 litres) of milk. Firstly the correct volume of rennet must be dripped carefully into a jug and then diluted with six times the volume of water. The mixture is well stirred and poured into the warm milk, and stirred slowly for a full three or four minutes. The lid is placed on the tub and the milk is left for around an hour to set. If the weather is cold wrapping a blanket around the tub will help to hold the warmth as the curd sets.

While the curd is setting the moulds can be prepared. These are made of elm wood, which resists the wet well and carries little flavour of its own. Each mould comes in two parts that sit one on top of the other. They are both open-bottomed rectangles around 7½in (19cm) long, 5in (12.5cm) wide and 6in (15cm) deep. The top one has a series of small holes drilled through its sides to help the whey escape.

A clean moulding board is set upon the bench and a clean straw mat is laid on top. The moulds rest on the mat. It is important that the mat is a little larger all around than the mould. It should be tidily trimmed, as the mat is sold with the cheese, supporting this very delicate product.

Once the curd is firmly set it is lifted gently with a skimming dish in whole 'leaves' that are laid into the moulds, reserving a particularly complete piece for the top layer. The mould is filled to the top.

No pressure is applied at any time; the whey is permitted to drain away of its own accord. Once the cheese has sunk below the top potion of the mould, this can be removed. The lower mould remains in place until the curd is firm enough to hold its shape. The top of the curd should curl inwards all around the edge.

The cheeses are now ready to be packed for market.

Devon Splits

Many people nowadays are under the impression that a Devon cream tea traditionally consists of jam, clotted cream and plain (or even fruit) scones. But scones are not a Devon custom at all; they were brought down from the Home Counties by the Edwardian summer visitor. A Devon cream tea was traditionally served with 'splits' or 'tuffs' or 'cut rounds' or even 'chudleighs'. They were all more akin to a soft light bread roll, not quite as rich as a scone. Different areas tended to prefer one name and recipe over the others, although they all were fairly similar. Splits, tuffs, cut rounds and chudleighs all allow you to pile on more jam and cream. They have the advantage of staying fresh for longer.

INGREDIENTS

1 lb (450 g) plain white flour

A pinch of salt

¼ pint (150 ml) water

A pinch of sugar

1 heaped tsp dried yeast or ½ oz
   (15 g) fresh yeast

6 tbsp (90 ml) full-fat milk

1 oz (25 g) lard

2 oz (50 g) butter

METHOD

Sift the flour and salt together into a large bowl and stand somewhere warm – I pop it on the far corner of the range for a few minutes. Next warm the water to blood heat. Once the water is warm add the pinch of sugar and your yeast. Leave it to stand for a few minutes for the yeast to begin working. Little bubbles will appear and you should be able to smell the yeast.

The milk, lard and butter go into a small saucepan. Heat them gently to melt the two fats into the milk. Be careful not to let it boil as this would scald the milk, changing both its texture and flavour. Once the fat has melted allow the mixture to cool for a few minutes. If the liquid is too hot when it comes into contact with the yeast it could kill it.

Take your warmed bowl of flour and make a well in the centre. Now pour in the milk and fat mix and check the temperature with your finger. It should feel just warm to the touch. Add the flour and mix everything together.

Sprinkle some flour onto your table and turn the dough out on it. Knead the dough well until it becomes elastic and springy. Return the dough to the bowl and cover it with a damp cloth. The bowl should now be stood somewhere warm for an hour or so – not on the range this time as that would be too hot. Yeast is happiest and grows fasted at body temperature. Avoid standing the bowl in a draught. As it stands it will begin to rise.

After about an hour – longer if the temperature is a bit on the cool side – turn the dough out on to the table and lightly knead it again.

Break the dough into around 15 to 20 pieces – depending on how large you like your splits – and form them into round rolls. Place them on a lightly greased baking tray about ½ inch (1cm) apart and return them to your warm place for their final rise. As soon as the splits have risen enough to be touching each other on the tray, they are ready for the oven. Bake them at 200°C or gas mark 6. I know the range is hot enough when it will lightly brown a piece of paper laid on the shelf for a few minutes. The splits take about 20 minutes to cook. When done they will be a nice gentle golden colour and sound hollow when tapped on the bottom with your knuckle.

To serve them, cut in half horizontally, spread on a generous helping of jam and then load up with clotted cream. You may find that your splits rise in the oven in such a way as to make it possible to tear them in half rather than cut them – that's why they are called splits.

In many Edwardian Devon homes people used dark treacle instead of jam on their splits and called the combination 'Thunder and Lightening'. I think I prefer jam, as long as it's home-made.

My first batch of tuffs ready for the oven. I experimented with both tuffs and cut rounds – finally settling on the tuffs for our consumption (they tasted the best) and the cut rounds for the tourists as they were so quick to produce.

The Edwardian Seaside Experience

Traditionally ladies and gentlemen were obliged to bathe on separate beaches but as the Edwardian period wore on, mixed bathing was increasingly found at seaside resorts.

I grew up in Bexhill-on-Sea on the Sussex coast in what is today a dormant retirement town. As I cycled the parades and wandered aimlessly along the promenades, it was clear from the architecture and layout of the town that it was once a thriving seaside resort. Hastings and Eastbourne, the two neighbouring coastal towns have managed to retain something of their seaside appeal and their pavilions, piers, winter gardens, theatres, parades and marinas all date from the late-Victorian and Edwardian heyday. I became fascinated with the former glory of what had become, in Bexhill's case, tatty windswept buildings and I was intrigued to imagine the many thousands of tourists who, at the turn of the century would have flocked to what was now a quiet and sleepy town.

Since the early 18th century the seaside, with the appeal of fresh sea air and salt water, had proved popular with the upper echelons of society. The therapeutic benefits of a short break on the coast were well recognized and, particularly during periods of antagonism between Britain and France, many seaside towns in England became the summer residences of the wealthiest of English society. The popularity of the seaside break grew steadily among the upper and middle classes. With the coming of the railways in the 1840s, more and more people were able to reach corners of our coastlines that had been difficult and costly to get to.

The birth of the true tradition of the British seaside resort probably came about with the expansion of the rail network in the 1860s and 1870s – subsequent cheap of travel opened up the coastline to the working-class populations of the biggest conurbations. By the Edwardian period this tradition had grown to its peak, with throngs of people of all classes making their way to the numerous popular resorts that peppered the coastlines of England, Scotland and Wales.

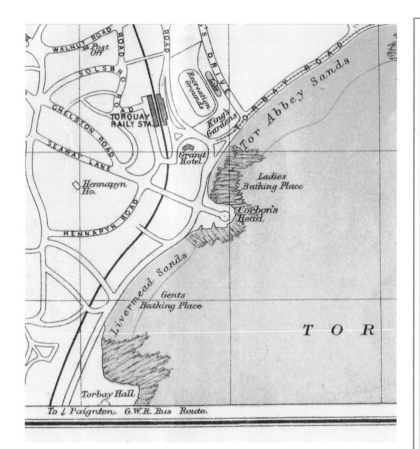

SEASIDE APPEAL

First and foremost was the appeal of the sea itself; an invigorating dip in salt water and a bracing walk along the coast were a refreshing alternative to dull city living, with its cramped housing conditions and dingy factory floors. The first bathing machines – huts on wheels that were drawn by horses out into the sea – are recorded from the 1730s; their modesty canopies protected the female bather from the 'gaze of idle or vulgar curiosity'. They became a stalwart of Victorian prudishness as they spared the blushes of the nation's female bathers and were to be found at resorts all round the coast. By the Edwardian period their use was on the wane as a radical liberalism took hold. Interestingly enough, it was my own now somewhat sedate home town of Bexhill-on-Sea that was first to allow, in 1901, bathers of both sexes to mix on the same beach. Beyond the attraction of the water, a whole host of entertainments grew up; in tandem with investment in accommodation, venues, facilities and municipal buildings, the seaside tourism tradition developed into the major coastline industry we recognize today.

ENTERTAINMENT

There was something for all ages and all tastes, from Punch and Judy shows through to penny peep shows and events such as galas, regattas and diving exhibitions helped to spice up the weekends. Stalls offered a rich array of souvenirs so that something could be taken away as a keepsake. Treacle apples, sticks of rock, china trinkets and glass ornaments were all commonplace – but by far and away the most popular souvenir was the picture postcard.

Much of all this we can still recognize as part of a visit to a seaside resort today and it is true that, really, things haven't changed that much. Tacky gifts, shabby food and barely passable entertainments remain a feature of the seaside scene and it is to the early years of the 20th century that we owe the tradition. By the Edwardian period, there were nearly three hundred established resorts in the United Kingdom and populations would frequently double and, in some instances treble, during the height of the season.

Accommodation

Firstly, of course, accommodation was needed for those who could afford more than a day trip. This began as lodgings in cottages and primitive 'guest' houses but steadily, over time, grew into purpose-built boarding houses and hotels. Nowhere more than at the seaside was the class-consciousness of the Edwardian period more keenly felt. Members of the refined upper classes were desperate to distance themselves from the boisterous antics of working-class tourists. Hotels with names such as 'The Grand', 'The Metropole' or 'The Imperial' gave refuge to the wealthier visitors, while working-class families shared rooms (sometimes even beds) and tables with fellow visitors in famously poor lodgings. For those who couldn't even afford the boarding house, a night under the stars on the beach or perhaps secretly stowing away in a bathing hut was all that could be hoped for.

The charabanc, derived from the French char à bancs *(meaning 'carriage with benches'), was used to transport tourists from the rural regions to local seaside resorts.*

PETER'S DIARY

I associate many a happy childhood memory with the towns of Lynton and Lynmouth in North Devon and experiencing them as an Edwardian holidaymaker was a fantastic day out. The coastline is stunning and I love the cliff railway that is powered by gravity: a water tank on the top car is filled and this weight pulls the lower car up the hill. (Until recently the ride down hill was free.) We enjoyed entertainment in the form of the only known troupe of Pierrotters operating in the country and a Punch and Judy show. We took a ride in a charabanc which was an early form of motorized transport and ate the standard seaside fare of fish and chips wrapped in newspaper (a tradition that ended with the alteration of the ink from edible to inedible). We also had an invigorating dip in the sea. I had made my own costume by painting red stripes onto my 'Victorian Farm' underwear which hadn't fully dried but I think looked fantastic. The paint took a long time to scrub off my skin but the memories will last a lifetime.

Spending Hard-earned Cash

Having sated their thirst for the sea, the hordes of tourists then became a captive audience to a whole range of entertainers, performers, tradesmen and tricksters waiting to fleece them of their spending money. Perhaps the defining characteristic of a seaside resort is a pier and, by the Edwardian period, piers had become the centre-pieces of the seaside experience. Starting out in the early days as a practical means of getting visitors who'd arrived by steamer boat ashore, they steadily grew in length and stature. At their peak, they become the locations for pavilions, theatres, side-shows, kiosks and amusement arcades. Entered through a turnstile from the promenade at a cost of between a penny and 'tuppence', the seaside pier was the place to see and be seen. With their gaudily embellished hand rails, balconies and benches and their beautifully decorative wrought-ironwork, piers encapsulated the exotic and escapist liberalism so prevalent in the Edwardian seaside experience. Oriental designs with domed and spiralled roofs mixed with the more traditional nautical themes such as mermaids, dolphins, anchors and mythical sea-serpents. The electric fairy lights that were fitted on Brighton's Palace Pier, which opened in 1899, must have created a truly magical experience for visitors as they paraded around in fancy and outlandish outfits.

The fun didn't stop beyond the pier. Ballrooms, tea rooms and concert halls provided entertainment throughout the day and into the night. Winter gardens – large wrought-iron glass-house structures mimicking the Great Exhibition of 1851 – provided a taste of the exotic for any winter visitors. Public gardens with water features, boating lakes and ornamental bridges offered an ordered retreat away from the bustling promenades. Up on the seafront musicians, acrobats, pedlars, photographers, fortune tellers, ice-cream vendors and donkey men all vied for custom.

This enormous popularity needed to be harnessed and a formalizing element took hold of resorts in the Edwardian age. Rules and regulations were drawn up, cafes and shops replaced itinerant tradesmen and mobile stalls. Licenses for entertainments were issued and acts confined to set locations in pavilions, theatres and concert halls. Order was introduced to the chaos of earlier times and the bazaar and fairground-like character of the mid-Victorian seaside scene was supplanted by a much more regulated and managed approach. It is then, perhaps, to the Edwardians that we owe the structure of the British seaside industry. Their legacy is the grand hotels, the ornate pavilions, the palace piers and the graceful promenades that still serve our seaside resorts so well.

ABOVE The infamous Mr Punch and his crocodile friend. LEFT An Edwardian troupe of Pierrotters provided fun and amusement for all the family in a traditional Edwardian seaside day out.

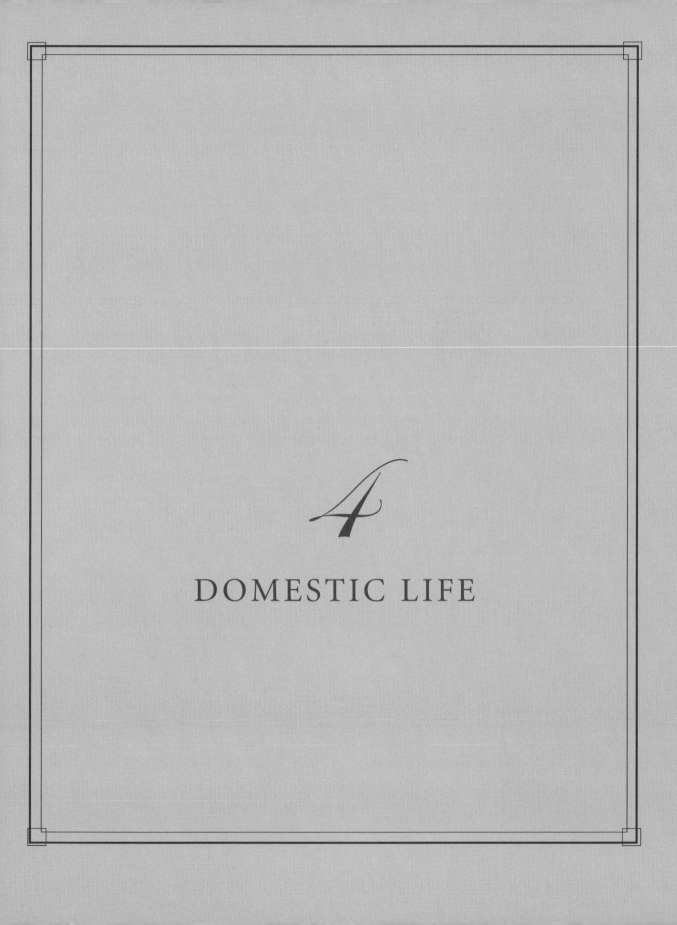

4

DOMESTIC LIFE

HEALTH AND HYGIENE

Mud is a fact of life on a farm. If you don't keep on top of it, it can make life miserable. In modern times the Wellington boot has been a great ally in the battle with mud. They allow you to leave most of the mud at the door when you slip them off. Although there were rubberized Wellington boots in the Edwardian period they were quite pricey, fitted items and their potential had not been noticed by the ordinary farming community. Just about everyone in the Edwardian countryside wore lace-up leather boots. They are warm and – if well looked after – waterproof. However, they also take ages to get on and off, and it is simply not practical to be forever lacing and unlacing them. So the mud was traipsed in all over the kitchen floor and, if you were not careful, all over the rest of the house.

Keeping an Edwardian home clean and bright was a continuous campaign of action. Coal dust, coal smuts and mud were constant companions.

Much of the daily cleaning regime at the cottage is about mud control. When the mud is dry it is fairly easy to sweep it up. Early in the morning is good for this as anything walked-in in the evening has had a chance to dry overnight. It's a good plan to sweep out the bedrooms and the staircase each morning after emptying the slop pails and before making the beds. The kitchen, however, requires a bit more attention. Since this is where most of the 'living' happens, it gets much dirtier than anywhere else, and of course this is where the food is prepared. So the kitchen floor needs regular scrubbing as well as sweeping.

The Edwardian period saw an explosion in the number of cleaning products available in the shops.

HOW TO SCRUB A FLOOR

Mrs Beeton in her *Book of Household Management* gives careful instructions for scrubbing floors. It may seem odd to write about something as simple as this, but doing it the right way saves a lot of work and makes the job much pleasanter.

First sweep the area thoroughly. Next assemble the equipment: two pails of water, a scrubbing brush and two cloths. For most dirty floors you will need no chemicals at all, just water. If, however, you think the floor has become very greasy – say someone spilt a hot dripping over it – then warm water with a bit of soap dissolved in it should replace one of the buckets of water. If you are worried about germs – maybe a baby regularly crawls on your floor – then a good splash of vinegar should be added to the water.

Start in one corner. Dip the brush into your first bucket, shake off any excess water, and scrub the area in front of you that you can comfortably reach. Take the first cloth and mop up all the dirt that the scrubbing has loosened. Rinse it out in the first bucket and wring out thoroughly. Next, dip the cloth in the second bucket of clean water, wring out and rub over the floor to rinse it off. Use the second cloth to dry the area. Now you can move on to the next patch of floor.

This method may seem long winded, but it works, and is much quicker than slopping water all over the place. It ensures that you, your clothes and the floor remain comfortably dry. It also does an extremely thorough job.

THE PRIVY

In towns, the Edwardian middle classes, and even some working-class households, were beginning to enjoy the benefits of the water closet. Rural populations, however, were largely without such a luxury. Mains water supplies and mains drainage took a long time to reach the countryside. Few villages were on the mains before the 1950s.

Our Edwardian farm came with a couple of dilapidated earth closets or privies. In Devon they were often built with an integral pigsty attached, as ours were. An earth closet relies on the waste breaking down naturally and composting. It works best when there is a good balance of different waste matter, which allows the right bacteria and micro organisms to do their job. If too much waste is added too quickly, or if the natural organisms are killed off, stench and disease are all too likely. If you have ever stayed somewhere where waste water empties into a septic tank rather than the mains drains, you will have seen instructions next to the loo about which chemicals and cleaning products can be used. If bleach is used, it kills the useful bacteria and micro organisms, and the tank goes haywire.

In modern Britain, there are also strict rules about the size of tank required for the number of people in the household. Overuse of an earth closet can cause serious health problems. In towns with a rapidly growing population, one privy designed to serve just one family suddenly became used by a dozen. So much waste in one place could not safely break down fast enough. Drains and sewers were the answer. By the beginning of Edward's reign huge civil engineering projects to install them were transforming the lives of town dwellers.

Back in the countryside, there are a number of basic management techniques that you can employ to ensure that using an earth closet is a reasonably pleasant experience. Firstly there is the site. You need drainage. The solid matter breaks down much quicker and in a much less smelly way if the liquids can drain away into the earth and if there is other organic matter present to aerate the mass. The drainage must be clear of everyone's water supplies.

If you have ever composted garden waste you will understand the principle. A heap of food waste and soft green weeds turns into a slimy smelly mess that takes ages to rot down. A compost heap that is all twigs and dry leaves also takes ages to decompose – although with rather less smell. But a mixture of the two works quickly and almost completely without any smell.

A Devon pigsty and privy arrangement works very well because of the straw that gets incorporated. When you are mucking out the pigs you discard soiled bedding at the same time as the dung, and this straw provides the other organic matter that the privy needs to function properly.

COAL

Coal was the other main enemy of Edwardian cottage cleanliness. Coal dust, coal ash and coal smuts were the price paid for heating, hot water and cooking. Every morning began with clearing out the range and black leading it. New coal was brought in, the fire lit and the dirt began again. For those not used to living with coal fires, it is the smuts that are hardest to understand – these fine, black, sticky flakes float in the air, gently settling on all surfaces.

It absorbs excess liquids and keeps the texture of the heap open, allowing oxygen in so that micro organisms thrive. Dry leaves are also an excellent privy improver – as is sawdust and scrunched up paper.

With the disposal side of the operation under control you can turn your attention to the privy's seat. The wooden seat should be kept scrupulously clean by daily scrubbing. The floor needs sweeping and scrubbing regularly and the walls should be whitewashed annually.

CLEANING PRODUCTS

With so much work involved in maintaining a basic level of cleanliness in a farm cottage, it comes as no surprise that many, and probably most, Edwardian farmhouses could be quite grubby when it came to the non-essential sorts of cleaning. Dirty windows and soot-stained walls are not a health hazard, so when time and energy are in short supply these sorts of things were often neglected. A good clear understanding of the difference between dirt dirt and germ dirt underpinned many a woman's household management.

Where time and labour permitted, most Edwardians aspired to the levels of cleanliness of the big house with its army of servants. As well as a good deal of household cleaning advice in popular books and magazines, cleaning products were beginning to creep on to the market.

Soap had been a boom industry in Victorian Britain and manufacturers were keen to continue their run of high profits. In the first decade of the 20th century soap came in a variety of forms. Soap flakes were marketed as ideal for laundry: essentially packets of grated soap, they allowed manufacturers to sell less soap for more money. Toilet soaps of many different types and perfume were sold for personal hygiene. Branded cleaning products appeared, based on basic recipes that people had been making at home for generations. But now for a little extra money you could buy it pre-mixed in a pretty packet. Chalk powder, for example, ready-mixed with water became the basis for several brands of 'cream cleaner'. Some added a little acetic acid. (You may like to have a look at what is in your own cream cleaner at home.) Metal polishes of various types were especially well represented.

Lower middle-class housewives tended to buy these new branded products. In the great houses of the day there was a strong skill base among the servants. They knew a whole host of recipes and techniques for cleaning everything from marble fire surrounds to wallpaper and had no need of expensive ready-made products. Poor people, of course, could not afford anything beyond a little basic soap, washing soda and starch. If you had worked as a servant in a large house before marrying, there was a good

HOUSEHOLD HINTS

Leather chairs and sofas may be revived by first touching up any scuffed areas with a little matching coloured ink. When this is thoroughly dry, apply a thin coat of whipped egg white and sugar. Leave it to dry for a few minutes and then polish off with a clean, dry, brush.

Cut glass is best polished when clean and dry with a soft brush and a little chalk dust, taking care that the brush goes into all the cavities and flutings.

Flies and other insects can be kept from settling on pictures and mirrors by rinsing them over twice a year with a solution made from boiling three or four onions in a pint (600ml) of water.

Plastering my jacket in a fine powdery dust didn't seem like the best way to get it clean, but it worked!

chance that you were able to bring some of this knowledge and skill to bear on to cleaning your own house. But if you needed a little guidance, you could turn to one of the many cheap publications aimed at housewives.

One of my favourites is *The Best Way, a Book of Household Hints and Recipes*, published in 1907. It was compiled from pieces that had appeared in two magazines: *Woman's World* and *Cosy Corner*. Much of the daily grind is ignored – after all, everyone knew how to sweep a floor and scrub a privy seat – but the magazines were a mine of information about the care of luxury items, with information on how to clean ivory, marble-topped wash stands and leather sofas.

BEAUTY PRODUCTS

The number and range of beauty products available during the Edwardian period grew enormously. Only a few years previously the list had been rather restricted, but now there were not so much hundreds but thousands to choose from. Attitudes towards beauty products were also changing fast. It was increasingly acceptable to use a wide range of scents and hair treatments, and discussion of them had become much more open.

In the May 1901 issue of the magazine *Woman at Home*, writer Annie Swan recommended 15 different beauty products to her readers. One reader was advised to try Dr. Weber's Comozone to increase the amount of hair upon her scalp, and clear up any dandruff. Further down the page she suggested another reader with thin hair growth should try using Prophylacticon followed by Nash's Concentrated Egg Julep.

To get rid of unwanted hair Annie Swan told readers to try depilatory creams such as Hair Foe, to use a pumice stone or consider electrolysis – for which she is able to supply addresses for treatment in London and Liverpool, but not Glasgow.

Make up was now worn openly, not just in medical cases to cover scars. *Woman at Home* listed many brands:

'*The best foundation for powder is a little Dysaline Cream (V. Darsy, 54, Faubourg St. Honoré, Paris), rubbed in gently all over the face, the superfluity removed and a little Printaniere Powder (same address) dusted over. This will tone down a red complexion nicely, without making the skin greasy.*

'*You can improve your sallow looks in the evening by applying Mme. Cross' Beauty Cream (4s. 6d.; 70, Newman Street, W.) all over your face, except where the colour comes. If very pale a little of her liquid rouge would improve your looks, and make you appear younger. You will find both of these and some other beautifiers in the Beauty Box, which she sends for 5s. 6d.*'

The same Mme. Cross is interviewed in a later edition of the magazine. It is a very supportive article; I imagine Mme. Cross was delighted with it, as it reads more like an advertising testimonial than a true interview. Her full product range gets mentioned: Beauty Cream, Wrinkline, Developer Cream, Anti-Corpo, Goldena, Egg Julep and Elderflower Water, and her premises and business methods are praised.

But as can be seen from the prices above, these products were well out of the reach of ordinary women. Instead, the home beauty tips in this magazine and other sources were much more practical.

Washing the face in milk was a long-established practice that helped to moisturize the skin. Elderflower water too was a tried and tested face wash. When the flowers are in season you can make an excellent moisturizer simply by putting lots of flower heads in a bowl and pouring a kettle of boiling water over them. Stir it about briskly and wash in the water as soon as it has cooled sufficiently. The flowers release natural glycerine and you can use the flower heads as a scrub. No product, no matter how much it costs, beats this simple recipe. During the rest of the year distilled elderflower water is fairly good, but the fresh stuff in season is best – an annual treat for the skin.

ALEX'S DIARY

In preparation for chapel, Ruth told me to scrub up a little so as not to let the side down. Usually when I'm out and about alongside Peter, I can get away with a quick brush down but it turned out that he had commissioned a new suit for the occasion. So I was going to have to do something if I wasn't to look like the poor relation. I'm always loath to get anything too wet down in this part of the world as trying to dry it in the damp climate can be a nightmare, so Ruth recommended a little book entitled 'The Best Way' – a collection of handy household recipes and tips sent in by readers of an Edwardian weekly magazine.

In it, there was a tip for cleaning clothes, particularly heavy woollen garments. At first sight it seemed illogical, as it involved smearing pipe clay (also known as Fuller's Earth and used as cat litter today) all over the jacket. Despite this seeming to be the exact reverse of what I'd intended to do, I set about methodically rubbing the fine powder into my jacket. I then folded it up and, following the instructions in the book, gave it a thorough thrashing with a flexible hazel wand. Then I unfolded the jacket and started brushing the clay off. To my amazement it had actually worked.

An excellent and practical book The Best Way *had an answer to most domestic problems.*

HAIR

The perfect Edwardian woman had long, brown, slightly curly, thick hair – and she had a lot of it. It was swept loosely up into a small bun on the top of her head, in a style that surrounded her face in a sort of halo of hair. There was remarkably little variation on the theme: photographs as well as fashion plates tell the same story throughout all levels of society. Working in factories or on the land, as domestic servants or strolling on the prom, almost every woman conformed. Occasionally you come across a picture of an elderly woman who has stuck to an older flatter style, but it is by no means common.

Big Hair

Not everyone has long, thick, curly, brown hair. My own while long is very fine and straight. When I sweep it up and pin it on top it immediately goes flat against my skull. I do not have 'Big Hair'. This was obviously a problem for many Edwardian women as can be seen by the number of products on the market claiming to thicken the hair and increase its volume. Prophylacticon, Mason's Jaborandi and Koko were all to be applied

With two plaited hairpieces coiled around my head, I could comb my own hair over into a suitable voluminous shape.

to the hair, while a new range of pills were available to promote hair growth. Capsuloids were to be taken three times a day before meals; the box claimed that: 'Capsuloids not only cause the death of those harmful germs which we have proved to be the cause of falling out and prematurely grey hair, but they also restore the injured growing cells of the injured roots, and nourish them, and cause them to multiply so that the roots become firm and grow rapidly, producing thick luxuriant hair.'

I try hard not to think about wearing someone else's hair on my head. It does get hot and itchy at times, but at least, being real hair, I can wash it as I do my own.

Washing and brushing regimes were believed to make a big difference to the condition and quantity of hair. Most advice centred around the correct water temperature and how often you washed it. 'Wash any thin patches daily in hot water' was one recommendation in a magazine; elsewhere in the same publication was the exhortation to wash the hair only once a month using a mixture of warm water, soap and ammonia. The length of the bristles on your hair brush was also thought to be critical.

When all the pills, lotions, potions and regimes failed to have any effect, women turned to hair pieces. It is possible to make your own. If you collect all the hair from your hairbrush, you can form it into rolls and pads to bulk out a hairstyle, but you need a surprising amount to make pads that are firm enough.

I use two long firm plaits of hair to pad out my own. They are both a mid-brown in colour, but the texture of the hair suggests that it was once much darker and has been bleached a little to make it more suitable for the English market.

I pin both the plaits around my head in a ring and carefully lay my own hair over the top, twisting the ends into a bun on the top of my head. Twenty or thirty pins hold everything in place.

THE HAIR INDUSTRY

The hair industry boomed during the Edwardian years. Some hair came from poor British women who sold it for cash, but as caps – which would have covered up their lack – had gone completely out of vogue, fewer and fewer of even the poorest women were willing to do this. Some hair came illegally from corpses – undertakers shaving the head just before nailing down the coffin. A lot of hair was imported from India, where it came mostly from the poor and desperate, although some was cut off as part of a religious ceremony.

Hair products became big sellers in Edwardian Britain, with hundreds of brands on offer.

SHAVE AND A HAIRCUT

For the filming of *Edwardian Farm* I decided to cultivate a full beard and let my hair grow (which is always a bonus in covering the bald patches), so that I could kick the series off with a shave and a haircut. Barbers are historically renowned as being only too willing to undertake a minor surgical procedure – a tooth removal or a bit of bloodletting as advertised by their red and white pole. So it was with some trepidation that I took to the chair for my first ever wet shave. An experience that I now believe every man should undertake at some point in his life. Walid Rostom who came from a long line of Lebanese barbers was a skilled man with a razor as well as a pair of scissors. The cut that he gave me made me look deceptively sharp, along with a collection of clothes that I had been given by Alex or found in the cupboards around Morwellham Quay. I was clean – a state that wouldn't last long working as we do day in day out on our farm.

ALEX'S DIARY

I don't think I had ever seen Peter so well-kempt in all my life (with the exception of his wedding day of course). His normally coarse and bristly cheeks were as smooth as a baby's backside and his hair was stunningly arranged in the classic 'short back and sides' cut, with a majestic centre parting. For the rest of the day I kept having to double-take as I observed this extremely well-groomed individual regally pacing around the farm. Anyone would think his mother was due to visit. It didn't take long, though, for standards to slip and within days the familiar five o'clock shadow began to reappear – occasionally developing into a full-grown beard but never quite receding to the neat shave he'd had upon our arrival back in September.

For my part, it took me a while to get to grips with a cut-throat razor.

Firstly, I had real problems getting it sharp enough to cut smoothly. I used a strop – a strap of leather with a link at one end to hang on the door – I ran the blade of the razor up and down it to get an edge on to it. To the strop I applied carborundum – a fine silicon carbide powder, first mass-produced in 1893.

ALEX'S DIARY

This is an abrasive and helped to get that razor-sharp finish. It took some nerve to cut into the bristles around my Adam's apple and a while for me to familiarize myself with the contours of my own face. After the first few shaves I looked rather like I'd had a fight with an extremely aggressive cat, but gradually my technique improved and shaving became less of a brush with death and more of a regular wash-time experience.

LEFT AND BELOW My first ever wet shave by someone else using a cut throat razor; a great experience once one gets over the initial trepidation.

Shaving by its very nature is something that man has had to contend with throughout human existence. There have been many fans of shaving over the years, including Alexander the Great and Beau Brummell. Shaving changed in a major way at the start of the 20th century with the patenting of the Gillette safety razor featuring a double-edged disposable blade. This idea had been toyed with by Gillette since 1895 but it was only in 1901 that it managed to get off the ground. The popularity of the safety razor exploded, culminating in Gillette supplying the US armed forces with disposable razors as part of their standard issue gear when they deployed to Europe to engage in the Great War.

Incidentally it is also at the turn of the century when the tune is thought to have originated, which is now referred to as 'a shave and a haircut' to which the response is 'two bits'. I will not tell you what the words are to this tune in Mexico, but it would make a sailor blush.

THE COTTAGE HOSPITAL

Before the advent of the NHS all medical treatment had to be paid for. Some medical insurance schemes were in place in parts of the country by 1900 and these helped with costs for those who could afford the premiums, but major emergencies could stretch the resources of even the wealthy.

Charitable institutions filled some, but by no means all, of the gaps. If Alex, Peter or I had suffered a medical emergency in Edwardian Devon we would probably have had to struggle through with very little professional help. Some doctors offered cheaper rates for the poorer members of society, and some did an element of completely free work. Drugs, however, still had to be paid for. Long-term illness could be the hardest, draining family resources and turning the sick person into a very real burden upon their nearest and dearest. Free hospitals funded by charitable donations did sterling work, but their scope was limited.

The Cottage Hospital in Tavistock was one such institution. It was intended to provide medical help and care for respectable working-class people who had been seriously injured. It would not take any infectious cases, so sufferers from tuberculosis, cholera or a thousand other conditions, had to look elsewhere for help. An injured

person could be admitted only if they had a letter of recommendation from one of the subscribers – the people who funded the hospital. (In practice this was always forthcoming if a doctor asked for it.) Once in hospital the patient had to make at least some contribution to the costs of their care. This was means tested and could vary from 3 shillings a week to 10 shillings and sixpence. Most patients spent around a month or more in the hospital. In 1901 Tavistock Cottage Hospital admitted 126 patients; this number roughly doubled in 1903, when a major extension was built to accommodate an extra 25 beds.

The hospital had a good reputation throughout the area, offering a high standard of medical care. If Peter had, for example, fallen from the top of the hay rick and broken several bones, the cottage hospital could have provided the very latest available treatment at a price that we could probably have managed.

The Widger family in 1907 took advantage of another free hospital for their son Tommy. He suffered from a double squint that threatened to prevent him entering either the army or the navy. Surgeons at the Plymouth Eye Infirmary offered to put it right. Treatment was free, though the family were still worried about how the lad would get on in the six weeks that he would be away from home. The operation proved a success and Tommy returned home at the allotted time – cured, if rather subdued.

It was remarkable quite how good the care offered in these charitable hospitals was, being in no way inferior to fee-paying institutions in most cases. But there were gaps. Some areas were served better than others; some types of medical problem got more attention and some less. The system was patchy – excellent in places and non-existent in others.

While more and more medicines were cheaply available at the pharmacists shop, many people still made up their preferred mixtures at home. I often had to improvise in our minimally equipped cottage. Smashing the sugar candy with the iron was enormous fun, somewhat spoiled by having to clean it afterwards.

TRADITIONAL COUGH REMEDY

INGREDIENTS
One stick of liquorish
The juice and zest of two lemons
2 oz (50 g) barley sugar candy
1 oz (25 g) currants
2 tblsp olive oil
1 tblsp brandy

METHOD
First grate the zest from both of the lemons and put to one side.
Take your liquorish root and bruise it as firmly as you can. I just kept bashing at it with my iron until it resembles a fibrous mat. Next squeeze the juice of your lemon into a small bowl and add the liquorish to infuse over a very low heat.
Now break up your sugar candy. I also used my iron and made a right mess. If you were to wrap the candy in a cloth or paper bag first and then beat it with a rolling pin you would achieve the same result without the mess. Put your candy into a second bowl and add the currants to it. Work the two firmly together breaking the currants as you so. Now add the oil and lemon zest, mix well.
Combine the two mixtures together and bring gently up to a simmer, mixing the whole time. As soon as all the sugar candy has completely melted take the mixture off the heat and add the brandy. Strain into a jar ready for use.
One teaspoonful to be taken after meals.

TEXTILES AND CLOTHING

CORSETS

The Edwardian period saw a dramatic change in the shape of corsets. A Victorian corset was a garment that controlled the waist, nipping it in and compressing the lower rib cage. It supported the bust, pushing it up. Below the waist, the Victorian corset was quite short, just covering the stomach at the front but stopping short of the hips at the side and back. There were a few tweaks in the design over the years; for example, to the busk, the stiff front panel of the corset, when spoon-shaped busks replaced straight ones. But the main changes were in fabrics, trims and stiffening materials: steel replaced whalebone in the 1860s and, around the same time, machine stitching replaced hand stitching. But for all the subtle changes, a corset was fundamentally the same garment doing the same job.

At the very turn of the century, corsets began to thrust the chest forwards and the hips back, creating an S shape to the spine and an hour-glass figure, a look defined by American illustrator Charles Dana Gibson in his drawings of women that became known as the Gibson Girls. Their images were reproduced everywhere – on china, tablecloths, fans and even wallpaper.

My corset is an important garment to me. It creates the right shape, and provides support and control.

Such corsets were by no means easy to make and their price remained high. Many an Edwardian woman made do with her old corset and just tried to alter her posture. The new drop-front blouses worn with a wide belt helped the illusion if she was careful how she stood.

Then the line of the corset began to lengthen down over the stomach and hips, and the bust was increasingly ignored. By the end of the decade the corset could have been more accurately called a girdle. Its new function was to smooth the waist rather than nip it in, and to flatten the stomach and reduce the hips. It provided only the most negligible support for the bust.

Changing Attitudes

Not only was the shape of the corset changing rapidly but so were attitudes. Throughout most of the Victorian period corseting was thought to provide a moral as well as a physical support to a woman. A 'neatened' waist would aid a woman to take pride in herself, to have greater self-confidence, to imbue her with the virtues of self-control and restraint. Only a slattern was loosely corseted. Tight lacing that unduly deformed the body was universally condemned, but the definition of what constituted 'tight lacing' was surprising. Reducing one's waist by three or four inches was considered perfectly normal and 'proportionate'.

The dress reform movement had regaled against the health aspects of corsets for decades. Much like the efforts of the modern Green movement, the first people to talk about the evils of corsets were seen as cranks. But towards the end of the 19th century a change began. More and more people addressed the issue from different standpoints. The artistic dress movement considered a looser line to be more beautiful; the medical profession was producing volumes of information on the dangers of corseting; while in Germany Dr Jaeger was promoting sanitary clothing that emphasized the need for the skin to breath freely (see page 177). (Today's well-known clothing brand still bears his name.) The debate about rational dress was one of the women's suffrage movement's themes, as well as being a topic in women's magazines and at church gatherings.

The next phase was one of commercial bandwagon jumping. Throughout the 1890s corset companies began to market 'healthy' corsets. In truth they differed hardly at all from the ordinary corsets – it was all window dressing, with a few perforations added here and there to help the skin 'breathe', or perhaps a small change in the layout of the boning. Those who were campaigning so hard to improve women's health must have been spitting nails to have their concerns hijacked in this way.

As the 20th century dawned the message was beginning to have the desired effect. In the health and personal appearance pages of a women's magazine in 1901, a letter was answered thus: 'I am surprised that any sensible woman should insist on her parlour maids being "well laced up from 8am till 10pm." Have you ever tried to do laborious manual work in tight corsets "well stiffened"? If so, I do not think you would insist on your servants wearing them.' Such a reply would have been unthinkable ten years earlier.

Jaeger's woollen corsets, stiffened only with lines of cording rather than steel or whalebone were one response, while the Knitted Corset Co. produced garments that included 'elastic bands at the side', offering a less-restrictive form of support.

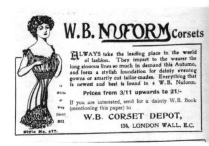

The new corsets began below the bust. In this advert, of around 1912, you can see that the bust is covered only by a lacy cotton camisole top, the boned corset encasing only the waist and hips.

Jaeger's own shops provided the very best quality woollen underwear, but there were many other cheaper brands around.

A bathing tent is probably the smallest confined space we have had to share in a long time...

Younger women began to experiment with less and less support in their corsets. I was told a family story of a young woman who, each time a piece of the boning in her corset poked through and began to annoy her, simply pulled it out and didn't replace it. Her mother was very concerned that she would 'ruin her figure', but the young woman didn't give a fig. Such stories became common in the Edwardian years as the taboos against unconstrained womanhood gradually broke down.

BATHING COSTUMES AND BATHING ETIQUETTE

The June 1901 issue of the women's magazine *Woman at Home* gives the following advice on bathing costumes: 'They should be made almost like loosely fitting combinations, with a gored skirt to fit over, reaching below the knee. It is well to make a bathing costume as pleasing to the eye as possible, trimming the black ones with broad red bands of braid and the dark blue with white. For children I should advocate the use of scarlet or white serge, in which they look so pretty! ... Children are never happier than when revelling in the sands with shovel and bucket; it would therefore be cruel to spoil their pleasure with the care of their clothes. The skirt, kilted or plain, and flannel blouse for girls and the serge sailor suit for boys are by far the most practical.'

The magazine also includes a fashion plate for bathing dresses. 'No.6 A serviceable and becoming bathing dress is a necessity for all who indulge in the healthful exercise of bathing. The example shown is in red and white galatea, strapped with bands of Turkey twill, stitched with white. The skirt is separated from the combinations, and fastens at the left side with pearl buttons.'

Men, meanwhile, were expected to wear something not so dissimilar. Male swimwear consisted of a pair of either serge or jersey knitted combinations in a dark colour, usually with short sleeves and trimmed in bands of white.

When the fashion for sea bathing had first taken off in the early Victorian period it had been seen mostly in terms of a health cure. But by the early 20th century people bathed mostly for fun rather than health. In the mid-19th century horse drawn bathing machines had carried the peaky bather through the breakers. They changed inside and then tentatively emerged on the steps. Outside, up to her waist in water – and fully clothed – waited the 'dipper' who had led the horse into the water and now rather unceremoniously dipped the reluctant bather. Five minutes was considered all that was to be healthful (except, of course, for the working-class 'dipper' who was wet to the waist all day). Then it was back into the bathing machine to get out of the wet bathing costume, towel themselves down and dress, while the 'dipper' manoeuvred the bathing machine back up the beach.

By the beginning of the 20th century there had been a few changes. For a start few bathing machines used horses any more – due to the problems with them dunging the water. Now most machines were hauled up and down on winches from the top of the beach. The 'dippers' too had gone; with bathing no longer a health procedure you could be as brave or cowardly as you wanted when it came to cold sea water. Bathing machines themselves were much less common. Many people were quite content with a small hut or tent on the beach where they could change, walking in and out of the water by themselves.

Photographs of beach scenes make it very clear that bathing costumes were only worn for actual bathing. On the beach people remained fully dressed. They might well be attired in looser and lighter-weight holiday clothes, but nonetheless they were dressed fit for the park or promenade. Children were permitted to take off their shoes and socks and hold up their skirts if they wished to paddle at the sea's edge.

The three of us emerged from the bathing tents with some trepidation.

FISHERMEN'S CLOTHES

It was easy to spot a fisherman on an Edwardian street by looking at his clothes. Chief among these badges of office was the Guernsey. Sometimes known as a Gansey and occasionally as a Knit Frock, the Guernsey was a knitted jumper well adapted for resisting all that the weather could throw at it. The style had been developed to perfection on the island of Guernsey in the 18th century and then spread among seagoing men all around Britain. It was so widespread and regularly worn, it almost qualified as a national costume.

In Devon and Cornwall the Guernsey was usually navy blue. It was hand knitted from very tightly spun four- or five-ply yarn which, by the Edwardian period, was almost exclusively produced in the mills of Yorkshire. The tightness of the spin made the yarn surprisingly water resistant. By knitting it up on fine thin needles and keeping the tension tight, the knitter could also make the Guernsey windproof. They were extremely hardwearing garments and were expected to last a man half a lifetime. They provided protection against the elements, while having no buttons or other parts that could get caught up in nets and lines.

The fishermen of Britain were instantly recognizable in their distinctive Guernseys and various caps.

Modern commentators often say that each village was said to have its own patterns that allowed a man to be recognized by his Guernsey alone should the need arise. But this isn't quite the impression you get from looking at Edwardian photographs. The fishermen of Devon and Cornwall were singled out by Edwardian amateur photographers as being especially picturesque and, as a result, we have plenty of wonderful images. The first thing you notice is the variety rather than uniformity. Each man's Guernsey is different. The men of one village, or indeed the men who man one boat, do not all share the same pattern. Even the Guernseys worn by two or three brothers are not identical – a knitter often found it sensible to make the family's Guernseys all different so that they could be told apart – though some patterns seem to have been more popular in some villages than others, and certainly some run in families. Nonetheless, a Guernsey knitted in Mousehole could easily be identical to one from Padstow.

In her book *Cornish Guernseys and Knit Frocks*, Mary Wright recorded a story from a family of Port Isaac:

RUTH'S DIARY

When I first saw the boys Guernsey jumpers I was immediately impressed and intrigued. They were beautiful true works of art and skill. I wanted to have a go myself, but how do you knit so tightly and evenly?

The answer turned out to be the knitting sheath. This is a stick that you can slip onto your waist band. It has a hold drilled in the other end that you can jam one of the knitting needles into as you are working. All the professional hand knitters used to use one; it not only improved the quality of their work, but also speeded it up.

I decided that I would have to learn. I was all fingers and thumbs to begin with. I dropped the dratted thing about every fourth stitch and gave myself neck ache holding my body all crooked and tense. But, it is beginning to come now, the rhythm is there and my speed is definitely increasing. I haven't knitted a Guernsey yet, but several pairs of mittens and a tam o'shanter hat further on, I intend to get started on a Guernsey next.

'Uncle Willie lost his, couldn't find it anywhere and Granny was mad. She had a stall every week at Rock and Padstow markets. Twelve months after, Granny saw a man wearing Uncle Willie's jersey. "Here," she said, "you' got my boy's jersey on." "I hab'n," he said. "Yes, you have," she said, and called a policeman to arrest him. "You make'n lift up his arms," said Granny. "You'll see I knitted a 'W' under one arm and an 'S' under the other and my boy's name is Willie Steer – what's his?"

A fisherman's hat was also distinctive. While the Edwardian landsman favoured the cloth cap or bowler, the fisherman was more often seen wearing the tam o'shanter or, if the weather looked wet, the sou'wester. Made of several layers of cotton stitched together, the brim of the sou'wester was sometimes stiffened with rows of cord trapped between the lines of stitching, just as women's sun bonnets were. Once sewn, the hat was saturated with linseed oil and allowed to dry. Once dry the hat became waterproof, but remained flexible. Warm woolly flannel linings could be worn underneath, either as a separate cap or sewn in. The fisherman could turn the brim up or down as it suited him and the weather. The deep brim at the back helped to

CONTRACT KNITTING

While many fishermen in Devon and Cornwall were lucky enough to have their Guernseys knitted for them by the women of their families, not all did. Contract knitting was an important source of income, especially in Cornwall where many thousands of women were employed on a piece-work basis producing Guernseys for sale all over the country. The yarn from Yorkshire was parcelled out by agents who later collected in the finished garments and paid the knitters. Within weeks a Guernsey knitted at Polperro might be on the back of a fisherman in Lowestoft.

funnel the water back over his shoulders rather than down his neck. At the front and sides the brim was a little narrower but could still be pulled down to offer protection to the ears and eyes.

Sou'westers were a simple garment to make, and many were knocked up at home, although you could also buy them easily enough. Whole coats of oiled cloth – oilskins – were also available, although many fishermen eschewed them as they were hard to work in.

Life jackets were even rarer – except among lifeboat crews. Within the fishing communities the life jacket was more a badge of pride than a practical garment. Men who manned the life boats liked to be photographed wearing their life jackets as a symbol of their position. The jackets themselves were hardwearing cotton tunics with large pieces of cork attached or sewn in. There were a number of different designs. In some the canvas was oiled like the sou'westers and oilskins, and in others not. Different systems of straps and buckles held them in place; some covered only the upper part of the chest and others extended down below the waist. The one thing they all had in common was the cork to give buoyancy. This made them quite pricey, which explains in part why there were so few aboard the *Titanic* – as well as so few among the ordinary fishing communities.

Life jackets and sou'wester's proved quick and easy to knock up at home.

DR JAEGER AND HIS SANITARY CLOTHING SYSTEM

A health regime based not on food or even exercise, but on clothing may seem a little strange. While Dr Gustav Jaeger, German physiologist and zoologist, believed that a proper diet eating and exercise were essential to good health, he felt that the topic of clothing had been neglected – with dire consequences.

The medical profession at this time believed that the skin needed to 'breathe'. Experiments on animals had shown that varnishing the skin to seal the pores led to death. Nowadays we would ascribe this to overheating. (When perspiration is prevented the body temperature can rise to dangerous levels.) But in the late-19th and early 20th century blocking the pores was thought to prevent a form of breathing. Current beliefs also held that some bodily waste products were disposed of through the pores of the skin, so blocking their action could lead to a poisonous build-up within the body. Mainstream medical advice was to keep the skin clean and free of any blockages.

Where Dr Jaeger's point of view differed was in believing that only clothing made from animal fibres allowed the healthy action of the skin. As far as he was concerned, all other fibres were unhealthy and insanitary. He devised a system of pure wool clothing, which he believed would promote general health, as well as curing a variety of skin diseases, firming the flesh, promoting better breathing, reducing infection rates from open wounds, helping the obese to slim down, and guarding against coughs and colds.

In his 1911 book *Health Culture* he described his ideal healthy wardrobe of pure woollen garments – from fine knitted combinations, to five-toed socks, woollen corsets, shirts and outer clothes. He extolled the virtues of doing away with sheets and sleeping only in wool, and recommended open breathable footwear, such as sandals worn over socks.

Dr Jaeger worked with manufacturers to produce his clothing designs and opened a chain of shops to sell them to the public. The business was very successful and his ideas were to have a huge impact on the way people dressed, especially among those interested in the healthy outdoor lifestyle so much in vogue in the Edwardian period. The Boy Scout and Girl Guide movements both paid close attention to Jaeger's concepts, as did many sporting organizations. Shackleton on his famous 1907 British Antarctic Expedition had the whole crew clothed by Jaeger, while in 1910 Scott headed in the same direction kitted out in Jaeger underwear.

When Roosevelt, by then ex-president of the USA, went to Africa he too wore the Jaeger range. Children were increasingly dressed according to 'healthy' principles, even when their parents conformed to conventional fashion. As these children grew up they often kept their preference for loose, healthy clothing – in turn driving fashions to change.

RAG RUGS

Few Edwardian homes were fully carpeted. Most people got by with oilcloth, linoleum, bare boards and a collection of rugs. While the wealthy might have Persian rugs, more ordinary people turned to homemade rugs.

Downstairs in the kitchen I find long hours standing on hard damp tiles something of a trial. My feet get cold, my legs start to ache and in the depths of winter I suffered from chilblains. Down came the rag rug, out of the bedroom, every time I knew I had a time-consuming job to do at the kitchen table. Wow, what a difference it made. I wish I'd had the time and resources to make several more rugs. I can see how sensible the Edwardian tradition was of making a new one every year, starting them off nice and clean and fluffy by the bed while older, more worn versions ended up in front of the kitchen range.

There are two basic types of rag rugs: those created by plaiting strips of rag together into one long continuous rope which is then coiled around, snail-like, and stitched in place; and those created by knotting strips of rag on to an open-weave fabric base.

Stepping out of bed first thing in the morning in an unheated house is chilly – and bare feet on bare floorboards doesn't help. The bedside rug is one of Edwardian life's mini luxuries.

I made the knotted rag rug. First I unpicked an old sack and laid it flat. Sacks are ideal: not only are they exceedingly cheap but the loose open weave is exactly the right size to pass a strip of rag through. Next all my rags had to be cut or torn into roughly equal strips.

I folded each strip of rag in half lengthways and pulled it partially through the sacking with a hook. A crochet hook would have done, but I found myself a proper rugging hook. These were sold in haberdashery stores up and down the length of Edwardian Britain. Having pulled the loop through the sacking, you pass the hook back through the same hole and use it to pull the two loose ends through both sacking and rag loop.

I loved doing it, but I have to admit it did take an inordinately long time to make my rug. When I worked it out, it was around 50 hours work. Looking at the size of the holes and rag strips, I had expected it to be a quick job, just a couple of evenings. Oh, no. To get a nice fluffy rug you need good coverage – which means that the strips have to go in close to each other. It all adds up to a lot of strips and a lot of hooking. So it was just as well that I enjoyed it.

Another surprise was just how much fabric it took. I estimate I used between three and four blanket-sized pieces of fabric to make a rug only four foot by three foot (1.2 by 1 metre). I could have made it less thick and shaggy, which would have economized somewhat, but whichever way you look at it, a rug is still a rag-hungry beast. Having used every scrap of fabric from my previously overflowing mending box and having sacrificed two worn blankets, I had to resort to cadging further fabric from everyone around. A worn-out pair of trousers made a six-inch (15cm) circle in one corner while somebody's old curtains gave me a four-inch (10cm) wide strip across the bottom.

I do wonder how poorer people could have found enough rag. In weaving areas, waste from the mills probably made rugging something that even the poorest could do, but elsewhere I have my doubts. The truly poor in Edwardian Britain had few rags to spare: adult clothes that fell apart were mended and cut down for the children, and scraps became dishcloths. Many a family is recorded as having only a single blanket on the bed, with every textile in the house being piled on top to keep out the cold. Adverts of the time portray rag-rugging as a nice lower-middle-class activity.

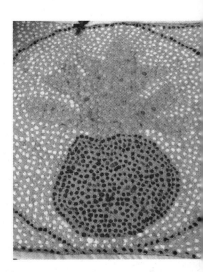

The design was much more discernable on the back of the rug, perhaps I should have gone for something much simpler. One of the advantages of rugs is how easy they are to get clean. A few hearty thwacks with the carpet beater and all the dirt flies out – more more hygenic than a fitted carpet.

A Woman's Place

For more than a generation the political argument over whether women should be allowed to vote had been raging. The very first time Parliament had been asked to discuss the issue was way back in 1832. The next came in 1867, presented by philosopher and social reformer John Stuart Mill, who was MP for Westminster. By the late 1890s a head of steam was building.

We sometimes forget that the same period saw a fight for wider male representation in Parliament as well. Nor is equality just about political power – or lack of it. Equality of education, equality of healthcare provision, equality of opportunity, equality of pay were all on the agenda – as they still are. The Edwardian period may not have seen the start of these ideas and it certainly did not see the end of them, but there can be no doubt that this is when the battle flared fiercest.

More than 35 political groups formed to promote women's suffrage, and several more to campaign against it. Many had extensive networks of local branches scattered throughout the country. In 1905 the struggle turned violent and arrests followed.

Those campaigning for women's suffrage took a number of different standpoints. Frances Cobbe, writer, social reformer and suffragette, believed that votes for women would bring about a moral improvement in society. Women she believed, while weaker in both body and intellect than men, were more sympathetic, practical, religious and moral – and their involvement in the political process would bring these qualities into government to benefit the nation in general.

When most of us think of equality, the suffragettes of the Edwardian period spring to mind.

For Helen Taylor, John Stuart Mill's step-daughter and suffragette, it was a matter of human rights; the 'right to belong to herself'. A strong sense of the injustice of the status quo was widespread among those who spoke up. Political power, it was hoped, would allow many inequalities to be finally addressed: unfair divorce law, child custody law, 'white slavery' and employment conditions were often cited.

RECOGNIZING WOMEN'S WORK

But what impact did politics have on the day-to-day lives of ordinary people far from the centres of power?

From the 1860s onward there had been a long and sustained movement towards codifying, formalizing and professionalizing traditional aspects of women's work. Teaching and nursing were at the forefront. Women were trained, their knowledge and competency tested, their status recognized and advertised. Most women in these fields embraced professionalization as it promised a higher status and recognition of their value to society, while patients and parents welcomed it as providing assurances of the quality of service.

As the 19th century progressed, a new set of milestones in the great equality race were reached. Elizabeth Garrett qualified as the first female doctor to be recognized by male professionals in 1865. In 1877 Annie Rogers passed all the necessary exams at Oxford University to gain a first-class degree, although they wouldn't give it to her – or any other woman student – until 1920. The following year London University became the first to admit women undergraduates on the same terms as men; Newnham College in Cambridge opened its doors to women students in 1880. Agnes Forbes Blackadder was the first woman to be allowed to graduate in Scotland in 1895, going on to become a doctor in 1901. There were, of course, many others, breaking the mould and raising expectations. By 1901 at the start of the Edwardian age, 15 per cent of all students in higher education were female, and by 1908 the first woman university professor, Edith Morley, was appointed.

Local Impact

Our local school in Morwellham was taught by a woman, Miss Mary James. She was a professional with a salary sufficient to live a respectable independent life. Over in Tavistock, as well as female teachers there were qualified female nurses. The cottage hospital boasted a matron, staff nurse and four probationers, as well as a cook, laundress, house-

CAMPAIGNERS AGAINST SUFFRAGE

Other people were actively campaigning against female suffrage, many of them women. For them it was largely to do with biology. Men and women were made differently; they should have different spheres rather than apeing the roles of the other. This is from an open letter signed by a range of influential men and women: 'While desiring the fullest possible development of the power, energies, and education of women, we believe that their work for the State, and their responsibilities towards it, must differ essentially from those of men… To men belong the struggle of debate and legislation in Parliament… women's direct participation is made impossible by the disabilities of sex, or by strong formations of custom and habit resting ultimately upon physical difference, which it is useless to contend… We are convinced that the pursuit of a mere outward equality with men is for women not only vain but – leads to a total misconception of women's true dignity and special mission…'

Whatever your views on the subject, in Edwardian Britain you could not be unaware of the argument.

maid and ward maid. There were no doctors on the permanent staff; the hospital was run by the matron, who was in her thirties. A look through Kelly's directory (a sort of yellow pages without phone numbers) shows a number of local businesses headed by women. They were mostly working within traditional female areas and represented about 5 per cent of all businesses. Emma Lucas ran a dairy, the sisters Millie and Emma Parker ran a dressmaking business, Mrs Ann Joll Rowe was post mistress and grocer, Mrs Jane Searle is listed as a beef retailer, and Mrs Ann Luxmore ran a general stores.

THE BATTLE FOR EQUALITY

Meanwhile, other battles for equality went on. Debates about the curriculum in elementary schools and how it differed for boys and girls grumbled along. Higher educational establishments were slowly opening their doors to women students. The first tentative calls for equal pay begin to be heard, especially among teachers. Demands were made for improved healthcare for women, especially those with young children. Women doctors were slowly gaining acceptance, as well as growing in numbers.

WORKING MOTHERS

Alongside this new world of more formally recognized work in the Edwardian years, was a renewed belief in the importance of the mother in the home. The two ideas may at first seem contradictory, but if it must be remembered, it was this latter strand that held sway in most people's lives.

Middle-class social reformers in the last two decades of the 19th century, who tried to understand the poverty they encountered, concluded that the mother held the central position within the family. They became convinced that where the mother was out at work the health and even survival of the children was in jeopardy. Poverty forced women into work and it was poverty that was damaging the children's health. Mrs Pankhurst herself believed that the vote would allow women to change the country in such a way as to protect poorer married women from the need to work. 'When women have the vote they will see that mothers can stay at home and care for their children. You men have made it impossible for these mothers to do that.'

Much social reform carried with it a determination to penalize the working married mother. Reformers were keen to push her back into the home as way of improving the lives of her children. The Social Union of Dundee was quite explicit about this. They ran nursing mothers' restaurants that provided free meals to mothers of infants under three months old. 'One object of the restaurant is to discourage married women's work. At three of the restaurants it is the rule that no free dinners should be granted to mothers who have returned to their work.'

A single woman could strive to make a way for herself in the male world; she could pursue an education and, if she worked extraordinarily hard, undertake prestigious and socially applauded work in areas that had been previously closed to her. But once she married marriage all her energies were to turn from that world to the domestic. Her new focus was to be her children.

In practical everyday terms, the suffrage movement probably didn't seem to make much difference to most men and women. Those who took part in the rallies, who formally joined the various organizations and called themselves suffragettes were mostly urban, upper middle-class women. Ordinary working women rarely appeared among them. But that doesn't mean that they took no part in or were not affected by the campaign for votes for women. Alice Kedge was a maid of all work in London.

'After I'd listened to a woman standing on a street corner somewhere in Camden Town, saying why we needed the vote, I became really interested. I went to several meetings and I bought myself a "Votes for Women" badge. Tin it was, with a safety pin attached so you could fasten it on to your dress or coat.

'When my mother saw the badge she said: "What are you doing with that thing? Take it off this minute and throw it away."

"Won't you let me help you John?"

Many of the older generation were like her. They saw life as a never ending struggle, particularly for women, which wasn't ever going to change so you just had to put up with it the way they had. Well, I didn't throw away my badge but I have to admit I tucked it under the lapel of my coat because I didn't want to upset my mother and I couldn't afford to lose my job. Even if nobody else could see it I knew it was there! I wasn't a brave soldier in the women's army but what the movement did was to make me think and then to believe I could do things for myself. And I did! People like me didn't write books about their experiences, so the effect gets forgotten. The suffragettes certainly helped to change my life.'

The world wasn't going to be the same again. Half the population had been given the go ahead to raise their expectations. The 1920s saw a frantic attempt to suppress it all after the Great War. It didn't work – the genie was out of the bottle.

Devon farmer's wives may not have been at the forefront of the Suffragette movement, but with the arguments raging in the newspapers and at public meetings up and down the country, they would certainly have had an opinion on the subject.

A Day in the Life of an Edwardian Farmer's Wife

As a woman with a home to maintain and with few machines and no servants to help, a portion of the day had to be set aside to keep things ticking over. But with no children to look after, my daily routine had a degree of flexibility that many Edwardian mothers could only have dreamed of. In general, on a non-washing day in winter, I could find about an hour before the midday meal and about four hours in the afternoon before supper, in which to do something other than the basic routine. Laundry completely filled that time on Mondays and often much of Wednesdays too, when the ironing was done. The summer was different as it was so much lighter in the evening. Work days were longer and it was possible to get more done. Here, a Thursday in February:

5.45 a.m. – Out of bed.
I dress by candlelight. I don't bother with my hair yet.

6.00 a.m. – Apron on and start cleaning out the range.
To work properly a range needs to be kept free from ash and soot. The only time you can do this is when the range is unlit and cold.

6.20 a.m. – Range lit and the kettle on.
I fill both my large pans with water from the bucket and put them on the range as well. The first pale glimmerings of dawn in the sky. I turn my attention to the kitchen floor. Like the range this is in constant use during the day, so I find it easiest to clean it now.

6.30 a.m. – The first warm water is ready.
I pour it into a jug and take it upstairs and bang on the boys' bedroom doors. They will be able to get up and have a quick wash before going out to see to the animals.

7.00 a.m. – More coal on the fire.
By the time I've scrubbed the floor, the boys have usually clattered down and gone out. Leaving the floor to finish drying, I take the second lot of hot water upstairs to have a wash myself. Despite all the technical advances in plumbing, most Edwardian women still washed with a simple jug and bowl system. After washing I put my hair up for the day.

7.30 a.m. – Back downstairs.
I give the table a quick rub over and start the breakfast. Tattie Raw Fry today. While breakfast is cooking I go out to fetch a couple of buckets of water.

8.45 a.m. Breakfast cleared away

I go back upstairs. In the summer I open the windows and throw back all the bed clothes to air, but on a cold February morning I just make them. I use any clean water that may be left in the jugs to give the washing basins and wash stands a wipe round. Now it's time to deal with the chamber pots – not my favourite job. I empty them into the slops pails and take the pails out to the privy before I scrub them and the chamber pots with washing soda.

10.00 a.m. – Clean the privy.

Devon privies come with integral pigsties. So I combine cleaning the privy with looking after our pig. She's a good-natured thing who likes a bit of fuss, so I usually spend five minutes giving her ears a scratch before turning her out so I can clean her sty.

11.00 a.m. – It's time to get the dinner on.

Some days I do a big batch of cooking that can be made to last several meals, as this frees me up to do other things without having to run back twice a day to cook. Many Edwardian women had a weekly routine: Monday to wash; Tuesday to dry; Wednesday to iron; Thursday to

bake; Friday to clean; Saturday to finish things up – and Sunday was another big cooking day. With the dinner on the way I now have that precious first free hour. Today it's a quick trip to the shop.

12.30 p.m. – We eat dinner.

2.00 p.m. – I get on with some paid work.

I am experimenting with the sorts of work that married woman did in this area. Gloving was one. Despite the huge technical advances in machines to make gloves, there remained a small market for hand-sewn gloves.

6.00 p.m. – Bring in more water and put a couple of large pans on the range.

Fill up everything ready for morning. I bring in the kindling and put it near range to dry out. The boys come in for supper.

7.00 p.m. – Just a few small jobs remain.

Things like the darning wait for the evening – it's good to have an excuse to sit down in front of the fire.

9.00 p.m. – Comb out hair, undress, go to bed.

Sleep like a log!

Food and Drink

 It's not just an army that marches upon its stomach. While much of our working life was taken up with producing Edwardian Britain's food, we also needed to keep ourselves fed.

COOKING ON THE FARM

Day to day cookery on our farm can best be described as cheap, filling and fast. There was little time for elaborate cookery, and few small-scale farming families had much money to spare for exotic ingredients. Our diet had to be based upon the basic supplies that were already available. People often imagine that farmers have always had plenty of access to a range of home-grown produce. Stories of wartime Britain often mention how farmers had far more access to the little luxuries such as extra eggs or plenty of milk and butter. These stories can be very misleading. Wartime Britain produced a very special set of circumstances, many of which worked strongly in the farmers' interests. The German U boat campaign was largely designed to stop supplies reaching Britain from overseas. Since the 1880s huge quantities of food had been imported, undercutting our own farmers while supplying cheap food to the urban majority. Grain came in from the USA and Canada, beef from Argentina, lamb from New Zealand. As those supplies were stopped, British farming rocketed out of depression in the scramble to feed everyone. Edwardian farming received no such boost. Farmers were struggling to make ends meet and the countryside was a much poorer place in general than the towns.

Those trying to make a living off the land needed to sell whatever they produced whenever possible. If you are living on the edge then it makes much more sense to sell

your half a dozen eggs and use the money to buy three loaves of bread. When times are hard you can live a lot longer on three loaves of bread than you can on six eggs. Those were the sorts of decisions that people had to make all the time.

Cheap food meant lots of starch, whether it was bread, potatoes, oatmeal or flour. Small amounts of meat and sugar helped to make the starchy foods more palatable. In the countryside we did at least have the benefit of vegetables and herbs, which were much more freely available than they were in towns. There were also the occasional treats. If an egg is cracked by the hen it is unsaleable and needs using straight away. Sometimes a glut on the market can depress prices, making it uneconomic to sell. And I don't think anyone has ever come up with a system that prevents strawberry pickers from eating their fill as they work.

Almost every meal we had on the farm included bread or potatoes as the main bulk in one form or another. Jam was another staple. Whether home-made or bought, jam was cheap and provided much needed energy quickly and easily. Imagine a world where there are no packets of crisps, sweets or chocolate bars that you can afford and you will begin to see how welcome a jam sandwich might be on a long hard day with several hours more work still to go.

The Local Shop

The local shop in Edwardian Britain had much more to offer in the way of food products than it had ever had before, with the mass manufacture and mass marketing of a whole range of items, but by modern standards, people ate very little in the way of branded goods. My most frequent purchases were bread, flour and tea. Oatmeal, lard and sugar probably came next. Soap, candles and then meat made up the third most common group. While this list was hardly different from the main purchases of someone a hundred years before, the packaging and origins of the certainly were. Flour in an Edwardian shop was most likely to come in a paper packet; the wheat it contained probably grown the other side of the Atlantic and milled almost anywhere in Britain before being delivered to the local shop. Many of the other products on the shelves told the same story. Even the most basic goods were no longer necessarily local. As a shopper I could have little idea of where things had come from.

In addition to the basics there were a few products that could be occasionally indulged in. Biscuits came in all shapes and sizes and at a range of prices. 'Family' biscuits were something that many working households could afford now and then, as were a range of ready-made sauces such as Worcestershire sauce, HP Sauce and Daddies Sauce. Both jam and marmalade were quite reasonably priced, as were syrup and treacle. Baking powder was something that was always on my shelf as it was both cheap and very useful in turning out quick easy meals. Ready-made suet and gelatine in packets represented a huge saving of time and labour and such 'convenience foods' quickly found their way into my shopping basket.

The local shop stocked a huge range of produce from candles to hoes, from treacle to potatoes. Surprisingly little of it was local produce, many of the foodstuffs on the shelves had come half way around the world.

ALEX'S DIARY

Throughout the year Ruth and Peter served up some of the most delicious and some of the most testing food I have ever experienced. It would be hard to pick my favourite dishes but Ruth's pasties (prepared under the watchful eyes of the local pasty aficionados) and Peter's luxurious carpetbag steak were real winners. Ruth's ginger beer flowed freely (albeit with a dash of rum) and Peter's potato wine certainly did the trick. The fresh fruits from the market garden slopes offered succulence and sweetness after a dry-salted and somewhat pickle-heavy winter and, all in all, I enjoyed the seasonal fare on offer. It would seem, however, that Ruth had finally found my gastronomic Achilles heel. I'm famous for eating whatever is put on the table and often brag about my strong stomach, but by serving a boiled sheep's head on a plate of stewed oats, Ruth had finally cracked me.

Just Out of Reach

Other products, although well represented in grocers shops, were outside our price range. Bars of chocolate sat tantalizingly in the window, along with tinned exotic fruit such as peaches. Tinned meat and fish really could not be justified on our budget – nor the increasingly large range of cleaning products.

Food in general was much more expensive in the Edwardian period; people had to spend a far higher proportion of their income on it than we do in modern Britain. Even the comfortably off were eager

not to waste anything. One of the things I did notice about the products available in the shop was how much smaller the packets were. Poorer people could only afford small quantities. If you can't store food you need to purchase it in small quantities daily. Food manufacturers were well aware of this and packaged accordingly. We were lucky enough to have a very serviceable pantry, even if it was damp.

As well as the economic circumstances of farmers and the new products available in the shops, our location had a huge impact on the food we ate. Our valley was subject to seasonal gluts of fruit, and some veg, tied in with the market gardening of the area.

A cold snap in late spring or early autumn could shorten the season, leaving surpluses on the market; while in a good year locals mostly did without.

While winter was a rather monotonous period for food, late spring and early summer saw an explosion of wonderful ingredients at our disposal. First came the rhubarb and the early gooseberries, then the madness of the strawberries, closely followed by raspberries, peas, salad stuffs and cherries, with runner beans following on. At much the same moment as the fruit gluts began, milk and cream production also came into full swing in the surrounding area. Had we been as little as 10 miles (16km) away, most of this produce would have passed us by. But in the Tamar Valley, soft fruit was cheap and plentiful. The best and earliest of course went straight to market, but towards the end of each fruit's season the price dropped dramatically. Early strawberries might bring in £40 an acre, but once the Lincolnshire growers came on line a few weeks later the strawberries were worth no more than £10 an acre for jam making. If you were a grower this was the time when you ate the most strawberries. It was a similar story for all the other fruits.

While our local shop stocked a bit of everything, the ships' chandler provided specialist items and equipment.

Meanwhile, the demand from the tourist industry kept up the prices for eggs, cream, butter and poultry. This was good for the producers but once again meant that locals had to wait for the prices to fall. Tourist seasons varied from year to year, partly in response to the weather.

Clotted cream was something that we had a very great deal of early in the milking season. The grass was lush so the milk fat levels were high, but the tourists had not yet arrived to eat the dairy produce. Cows have their own rhythm; if you want to milk them at all, you have to milk them regularly throughout the season – you can't start and stop as the demand for their produce fluctuates. So in order to supply the peak summer demand, farmers often had gluts at other times. Cheese making helps a little to even this out, but even so at certain times cream prices were very low.

Just as the resources of the region had their impact, so too did the customs. Devon and Cornwall both have very strong food traditions and we were bang slap in the middle, being only 10 yards (9 metres) from the border between the two. Local dishes reflected the ingredients available, but also had their own methods of cooking. The ingredients for a Cornish pasty, for example, are easily available in most parts of rural Britain, but combining them in this form is quintessentially West Country cooking. Many dairying areas of the country produce rich creamy milk, but clotted cream is special. Roast dinners locally are also still generally cooked with at least a cup full of water added to the baking tray, something that that Edwardian fisherman's wife Annie Widger was doing with her 'Baked Dinner' back in 1905. It's a practice that I have not come across elsewhere. Our valley made a speciality of the cherry pasty, combining the local glut with a more common form. Simple and delicious, the whole cherries – stones and all – were lightly sprinkled with sugar and wrapped up in pastry like a pasty. They were eaten with clotted cream and a warning to mind your teeth on the stones.

Clotted cream wasn't just eaten with sweet things like cherry pasties. Where most people in Britain might add a knob of butter to a dish, people in Edwardian Devon used clotted cream. I can heartily recommend clotted cream on baked potatoes, for example. It was also commonly used on meat puddings as a sauce. After all, Devon farmhouse butter was usually made by beating clotted cream with your hand until it separates into butter and a little buttermilk.

MAKING DO WITH THE BY CATCH

Fishing too had its impact on our diet. Many local men would turn to a bit of fishing in the winter when there was less to do on the land. In the Tamar Valley it was mostly salmon fishing during the Edwardian period – the fish rarely being eaten by the fishermen, but sold on to the more wealthy and those in the hotel trade. We were not able to fish the Tamar as it is currently not permitted while salmon stocks struggle to rebuild from the effects of pollution and overfishing in the past. Alex and Peter went a little further afield to try their hand at sea fishing instead. Once again the main catch went straight to market. It was the 'by catch' that ended up on our plates. These are the more weird and wonderful species that are harder to sell. A little bit of this and a little bit of that, which usually end up in a fish soup or pie.

SEASONAL FOODS

This is a table of the foods available to us throughout the year on our Edwardian farm. Here in Devon in a sheltered valley, the seasons are very different from other parts of the country. No single seasonal table would be accurate for all of rural Britain – like so many parts of this journey into Edwardian farming, this list is a reflection of just one experience of country life.

SEPTEMBER

Beans, carrots, lettuces, marrows, potatoes, turnips

Apples, damsons, pears

Goat's milk

Beef, chicken, lamb, mutton, pork, rabbit

DECEMBER

Cabbages, carrots, leeks, onions, potatoes, parsnips, sprouts, turnips

Beef, chicken, goose, mutton, pigeon, pork, rabbit, turkey

Herring, sprats

OCTOBER

Beetroot, cabbages, carrots, cauliflowers, marrows, onions, potatoes, turnips

Apples, pears

Goat's milk

Beef, mutton, pork, rabbit

JANUARY

Beetroot, cabbages, carrots, leeks, onions, parsnips, potatoes, sprouts, turnips

Beef, mutton, pork, rabbit

Herring, sprats

FEBRUARY

Cabbage, carrots, onions, parsnips, potatoes, sprouts, turnips

Beef, mutton, pork

Herring, sprats

NOVEMBER

Beetroot, cabbage, carrots, onions, potatoes, turnips

Apples

Beef, mutton, pigeon, pork, rabbit

MARCH

Cabbages, carrots, onions, parsnips, potatoes, turnips

Beef, mutton, pork

Herring, pilchards, sprats

Eggs

APRIL

Lettuces, parsnips, radishes, spinach, spring onions

Rhubarb

Butter, cream, eggs

Beef, chicken, lamb, mutton, rabbit

Pilchards

MAY

Beans, carrots, lettuces, new potatoes, peas, radishes, spinach

Gooseberries, rhubarb, strawberries

Butter, cream, eggs

Beef, chicken, goose, lamb, mutton, rabbit, veal

Herring, mackerel

JUNE

Beans, cabbages, carrots, lettuces, onions, peas, potatoes, radishes, spinach

Cherries, blackcurrants, raspberries, redcurrants, strawberries

Butter, cream, eggs, soft cheese

Beef, chicken, lamb, mutton, rabbit, veal

Herring, mackerel

JULY

Beans, cabbages, carrots, lettuces, marrows, onions, peas, radishes, turnips

Cherries, raspberries

Butter, cream, eggs, soft cheese

Beef, lamb, mutton, rabbit

Herring, mackerel

AUGUST

Beans, carrots, cabbages, cauliflowers, lettuces, marrows, onions, peas, potatoes, turnips

Plums

Butter, cream, eggs, soft cheese

Beef, lamb, mutton, pigeon, rabbit

WEST COUNTRY CIDER MAKING

Always keen to sample the local tipple and anticipating a hot summer, Peter and I jumped at the opportunity to make some traditional West Country cider. Up in Shropshire, we had used a horse-powered stone mill and a mobile double-screw press to make cider. This was very much a regional technique. In the south west, an altogether different method was used to produce what the locals termed liquid gold.

We headed across the Tamar into Cornwall to the great cider house at Cotehele manor and met with cider-maker Chris Groves.

The fantastic thing about cider making in Edwardian times was that it was so localized. Almost every farmhouse, manor house or smallholding made cider for their own consumption or for their labouring staff. Right up until the Truck Act of 1887 (a form of employment rights), farm labourers could legally receive their wages in cider – a deal that in many cases suited labourer and landowner alike. Large cider-making concerns like the system at Cotehele produced vast quantities to supplement the pay of the workforce for the entire estate.

We wanted to squeeze enough cider to fill a barrel – which traditionally held 31½ pre-imperial gallons (120 litres) – in the hope that it would be enough to slake the thirst of our harvest helpers. First the apples needed milling and for this Cotehele had a pony-powered rotary 'scratter' box. This ingenious contraption harnessed a pony to a horizontal wheel, which powered, via a gearing system, two iron-toothed rollers that pressed and mashed the apples.

Perhaps the most interesting part of the whole procedure was how Chris made the mats that separate the layers of mashed apple in the press. He simply laid long wheat straw so that it radiated out from the centre of the press in a square pattern, and set the pulped apples on top. The straw was folded over the top and secured by another handful of mashed apples. This arrangement is known as a 'cheese'. Another layer of straw was added to make another 'cheese' and so on, until we'd reached a satisfactory height. Then Chris

Cider and farming have gone hand in hand for centuries as the beverage was a way of providing safe drinking water and also a means of attracting an itinerant work force to come and see you through your harvest.

Peter and I steadily crank the head block down to increase the pressure on the milled apples.

climbed up on to the head block of the press and began lowering the top pressing plate. As it made contact a torrent of juice gushed down through a muslin filter sheet and into the large granite trough at the base of the press.

Pressing apples on this scale is a lengthy process and one that requires an inordinate amount of strength. Peter and I helped push on a large pole inserted into the screw mechanism to crank down the pressing plate. Finally, when we could push no harder, the pole was fastened to a fixed winch by a rope and this, in turn, was wound to exert yet more pressure. Still, the juice kept flowing. With presses of this size, the cider man would revisit the cider house every hour or so, to crank a little more pressure over the course of a couple of days.

Chris had been enormously generous with his knowledge and we thanked him gratefully. We were lucky enough to take the first 32 gallons (145 litres) of juice. As we rolled our barrel back through the orchard, we pondered how we were going to get it across the Tamar to Devon.

MANIPULATING CIDER

I was intrigued to learn there were many ways cider could be tinkered with to improve the taste. In pursuit of perfection, I was keen to explore some techniques to produce the perfect cider.

In the first few days in the barrel, if the temperature is warm, the cider should begin fermentation. It is not uncommon for a cider to get 'stuck', where fermentation has not kicked off – either due to a lack of yeast or, more likely, a lack of nutrients for the yeast to feed on. Should this happen, a handful of topsoil from the orchard is said to hurry along the process. The yeast can be 'fed' if it still doesn't appear to be working fast enough, by immersing a slab of bacon, a rabbit skin or a side of beef in the juice. Yeast works well on a diet rich in protein and apple juice can be lacking in amino acids and vitamins. I've even heard tales where a stuck cider has miraculously started fermenting in the middle of winter, only for the skeleton of a – presumably drowned – rat to be recovered from the empty barrel some five to six months later.

Our cider began fermenting vigorously a mere twenty-four hours after barrelling, so no tinkering required.

A dash of roasted beetroot juice or a couple of ounces of burnt sugar helps to darken up an otherwise pale drink to a reassuringly strong scrumpy shade.

High pectin levels in cider can lead to a cloudy and quite murky drink and it simply doesn't matter how long you let the barrel sit, it simply won't settle to a brilliantly translucent drink. To achieve this I would have to add some 'finings' to knock back the pectin, then I'd need to 'rack off' (using a syphon tube) the cleared drink. In modern cider and wine making, a pectic enzyme is added to break down the pectin, but traditionally bull's blood or egg white were used as finings to coagulate the pectin so that it sinks to the bottom of the barrel. However, all of this would have to wait for spring when the fermentation process was well and truly over.

A LESSON IN BUTCHERY

We decided to cash in early on our herd of Red Ruby cattle. Having sent the most promising bullock to slaughter, local butcher and huntsman Chris Rounsevell arrived to joint our share of the carcase. Peter and I were keen to give this famously fine-grained and marbled beef a proper taste test. After the carcase had been dressed neatly and cleanly, it had been left to hang for twenty-four hours. It was then cut down the chine or backbone into two sides. These sides were, in turn, cut in half (the whole carcase is cut into quarters), then hung for anything up to five weeks. The loins and the rump are, logically, found on the hindquarters, along with the buttock, hock, flank and tail. We all know the rump and sirloin are used as steaks and the topsides and silversides as roasting joints. The forequarters have, however, far fewer popularly recognized cuts and it was this part of the animal that Chris used to demonstrate how the meat was cut from the carcase.

Peter and I were a little dismayed to find out that the forequarter had no real parts that could be sliced fresh and fried like a steak. But Chris assured us that while the forequarter is not famed for its tenderness, it matches the hindquarter for taste. First there are the ribs. These break down into the fore, middle and chuck ribs and can, when de-boned, rolled and tied, make roasting joints. Once the shoulder blade has been removed from the middle rib, the spare ribs beneath can also be used as a roast or grilled dish. The neck, sticking and clod are the parts that run from the neck down to the belly

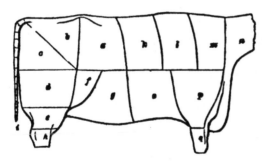

THE ENGLISH MODE OF CUTTING UP A CARCASS OF BEEF

especially in the fore-quarter. Fig. 278 shows this mode, and it consists of the following pieces :

In the hind-quarter.		In the fore-quarter.	
a,	The loin.	k,	The fore rib.
b,	.. rump.	l,	.. middle rib.
c,	.. aitch-bone.	m,	.. chuck rib.
d,	.. buttock.	n,	.. clod. and sticking, and neck.
e,	.. hock.	o,	.. brisket.
f,	.. thick flank.	p,	.. leg-of-mutton piece.
g,	.. thin flank.	q,	.. shin.
h,	.. shin.		
i,	.. tail.		
(1071)			

OPPOSITE: It was great to learn at last exactly where some of my favourite beef cuts come from. Much of the flesh from the forequarter would be diced for stewing steak.

or brisket of the animal. I diced these into a stewing steak while the brisket was rolled and tied. Nothing went to waste. The bones were boiled down for stock while the excess fat we cut off would be rendered down to tallow and used in a whole range of domestic and industrial processes. I took home the shin, which I was looking forward to cooking using Chris's recommended method – slowly pot-roasting it over a low heat for the best part of the day, with an assortment of winter vegetables and herbs.

The Red Ruby Devon is famed for its beautiful meat, marbled with fat and succulent taste.

CARPETBAG STEAK

Carpetbag steak is an Australian dish that I first came across in an international cookbook. I have read that the recipe had its origins in Victorian England with the nation's fondness for steak and oyster dishes (and some even say it is influenced by American cuisine). The directions that I followed were in an Edwardian cook book.

In its most basic form it is a beef steak with a 'pocket' made using a knife and stuffed with oysters (if oysters are not available mussels will do). It should be stitched up either using, as I did, a needle and canvas thread or by securing the opening with wooden toothpicks. The steak oyster combination is then seasoned and fried to your liking – rare, medium or well done. A little flair can be added to the dish with the ingredients such as mushrooms and red wine, or tomatoes and garlic.

I chose to cook this dish because, for me, carpetbag steak really sums up our project, with one ingredient coming from the land and one from the sea. Furthermore, many miners from the Tamar Valley emigrated to Australia to continue mining out there, so to cook an Antipodean dish seemed appropriate.

Food is forever evolving because it is essential to human existence. It is also a way for us to express individuality, regionality and culture. But, as the world gets smaller and food gets faster and more generic, we need to try to safeguard ingredients and dishes that may fall by the wayside. It will always be hard to ascertain the history of various dishes and the factors that have had an impact upon them, but right now the ownership of carpetbag steak is firmly down under.

PETER'S DIARY

With our football match only days away Alex and I realized that we would need something to drink at half time. Football sustenance during the Edwardian period was a long way away from the changing-room snacks of orange quarters and a packet of wine gums. More than likely we would have been feasting on a pint of ale and a cheeky fag (purely for health reasons). However, we both have some Scottish in us so we decided that beef tea made to an Edwardian recipe that we found in the 'Best Way', our household reference book, would make an appropriate nutritious half-time drink.

The 'Best Way' instructed us to take some beef, season it, boil it up and strain the liquid into a bowl. This should then be left to cool and consumed within a couple of days (heated prior to drinking if so desired). We were also advised that the remaining beef fibres would make a nice potted meat paste.

Liquid beef first fed the armies of Napoleon and still feeds the armies at the Old Firm Derby to this day. I fully expected the beef tea that we were making to be brown, but it was clear (or at the very most it had a slight green tinge). I am pleased to say that it was delicious and I would highly recommend it.

THE DAIRY AT LANHYDROCK

Lanhydrock in Cornwall boasts one of the finest dairy suites of any country house in Britain, capable of producing a great range of high-quality desserts for the table. Such vast spreads were an Edwardian sign of wealth. Those lower down the social scale tried to emulate it where they could by making jellies and blancmange, but few could produce the ice creams, iced puddings and sheer range of dairy items of a great house.

At Lanhydrock the ice room, dairy scullery and cool room are at the far end of the kitchen offices, kept separate behind a door that divides the 'hot' side of the kitchen from the 'cold'. Ice was stored in lead-lined safes in the ice room and ice-cream moulds lined the shelves, along with ice paddles and pails. Even when the house got its own generator, the ice cream was still made using only ice and salt, without the benefit of a freezer.

The cool storage room was the real showpiece, where creams, jellies, soft cheeses, butter and desserts were kept in perfect condition until wanted at the table. A wooden bench ran around three sides of the room, with a stream of cold water circulating slowly through a channel in the bench. A marble island in the middle of the room also incorporated a channel of cold running water. This was a way of keeping the room at a constant temperature and the air retained a great deal of moisture so that none of the foods dried out.

CLOTTED CREAM

Clotted cream is such an important part of the food culture in this part of the world that I felt that I needed to make an awful lot of it (and eat it, of course!). The true farmhouse clotted cream is made directly from fresh milk. At each milking, both morning and night, the milk straight from the cow is carefully strained several times through a fine cloth to remove any hair or dust that may have fallen into the pail. Once it is absolutely clean the milk is poured into a shallow bowl and allowed to stand overnight in a cool dairy.

The old farmhouse dairies of Devon and Cornwall are wonderful places. Always on the northern and eastern side of the farmhouse, they often have flagged floors and slate shelves. The walls are thickly whitewashed and small windows are placed so as to encourage a constant cool draught of air. I have seen some which route a small stream through the building in a channel down the centre of the dairy. This not only provides running water to keep everything spotlessly clean but also cools the space, an early and environmentally friendly form of refrigeration.

The pans of milk stand on the slate shelves as they cool from the cow. The milk rapidly falls in temperature from blood heat to an ideal temperature of around 39–46ºF (4–8ºC).

Mrs Beeton's Fruit Ice Cream

INGREDIENTS

To every pint (600 ml) of fruit juice allow

1 pint (600 ml) of cream

Sugar to taste

METHOD

Let the fruit be well ripened; pick it off the stalks, and put it into a large earthen pan. Stir it about with a wooden spoon, breaking it until it is thoroughly mashed; then, with the back of the spoon, rub it through a hair sieve. Sweeten it nicely with pounded sugar; whip the cream for a few minutes, add it to the fruit, and whisk the whole again for another 5 minutes. Put the mixture into a bowl, and freeze, taking care to stir the cream, etc., two or three times, and to remove it from the sides of the vessel, that the mixture may be equally frozen and smooth.

Strawberry, raspberry, redcurrant, whitecurrant or blackcurrant ice creams can all be made using this recipe, in fact, any fruit from mango to fig.

On the second day the pans of milk are ready for 'cooking'. The cream will have risen overnight and now forms a layer on top of the milk. As gently as possible – so as not to disturb the layer of cream, the pan is transferred to the fire. In the old farmhouses this would have been a wood fire, or up on Dartmoor a peat fire, and the smoke from these fires inevitably flavoured the cream. More modern establishments used a water bath cream stove, a wonderful example of which survives at Lanhydrock, where hot water is piped in from another room to swirl around underneath the pans – eliminating any fire and smoke flavours from the process. This was especially beneficial if your fuel was coal. While wood or peat smoke can be an acquired taste, coal smoke is simply nasty.

Whatever the heat source it was advisable to sit the pan in a larger bath of hot water so that the cream didn't 'catch' on the bottom. The water ideally should be just steaming and not yet quite at a simmer. It takes about half an hour for the cream to cook and requires careful watching throughout. It is ready when little bubbles appear around the edge of the pan and the cream is seen to 'lift'. The pan must be gently removed from the heat and returned to the dairy to cool. This time the pans sit not on the slate shelves but on a wooden rack that usually occupies one wall of a West Country dairy. Here it will cool more slowly and more evenly than it would on the slate.

Day three finishes the process. The cooked cream is carefully lifted off the milk with a skimmer and laid in a dish. The cream layer is quite thin in the settling pans, only three or four millimetres in unfavourable weather, more like a centimetre at most. When it is skimmed, therefore, the cream is laid one layer upon another, so that the finished product is many crusts thick, giving quite a different texture and indeed product to the cream most of us are familiar with.

More modern commercially minded clotted-cream producers in Edwardian Devon and Cornwall were employing a quite different method, one that is still widely used today.

Instead of beginning the process with the whole milk from the cow they first used a mechanical cream separator to take all the cream they could from the milk. This takes much more of the cream out of the milk than the old farmhouse method, leaving thoroughly skimmed milk behind. The resulting cream was then poured into a series of small shallow containers. These could be immediately plunged into the water bath for cooking without the overnight settling period. Once cooked they were cooled before being sold in those very same containers. This produced a cream that had only one layer of crust, on the top, with a deep layer of more ordinary cream beneath. It was, and is, a blander product but much easier to produce in volume.

THE DAIRY SCHOOL

Dairy Schools were a new idea that began in the last twenty years of Victoria's reign and were well into their stride by 1900. They were established all over Britain in rural areas to teach the 'new and improved' dairying methods to daughters of farming families. The schools held an important place in rural society as they reflected a new respect and professionalism for traditional female work. In many ways they represented the rural women workers' involvement in emancipation. Many women had worked hard to get a form of formal training and recognition for a range of traditional female roles. Nursing and teaching had been at the forefront of this push. Once women received training and passed equally rigorous exams as men, it became much easier to argue for equal respect, and improved pay.

A few women in Edwardian Britain went to the agricultural college but dairy schools were more widespread and more widely attended, as this was an area of agriculture that women were traditionally involved with. The new dairy schools focused on hygiene and scientific method within the dairy. This was partly to distance themselves from older informal practice and to associate themselves with modernity and professionalism – and partly an attempt to increase the chances of their pupils finding employment. A photograph from a dairy school at South Morton in Devon shows the young women using the latest in butter-making technology, with a cream separator, barrel churns and tabletop butter workers. This would have been radically different to the methods many

of them had used at home. While barrel churns had long been in use in Shropshire, for example, in Devon people had continued to use an ancient system, one that was already obsolete in the rest of England by the 16th century. A Devon farmer's wife clotted her cream over the fire then turned the finished cream into a shallow wooden tub and beat it with her hand until it separated out into butter and buttermilk. The resultant butter was rich and often carried a hint of wood or peat smoke. Those who had grown up with such flavours really enjoyed the taste, but customers in other parts of Britain found it unpleasant, which meant that there was no wider market for Devon butter, nor indeed one among the summer visitors. The new dairy school taught women to turn their milk and efforts to more profit.

Cheese making, too, was taught in a more scientific way, more in line with wider consumer tastes. Devon did not have a good reputation for cheese making, so if a dairywoman was to make a living she would need to be able to produce other more popular varieties and styles of cheese. Precise recipes were provided for cheeses that were proving saleable to the British public, along with expected prices at market and a careful analysis of the equipment and investment required. Two styles of pressed cheese and two styles of soft cheese were particularly favoured: a Gaerphilly-like cheese and a Cheshire-like cheese, along with one akin to a French Coulommier, and something termed Cambridge cheese. Brie and Camembert style cheeses were also achievable and saleable.

SPIRITUAL LIFE

The geography of religious buildings was laid out long before the break with Rome, before there was a Church of England, even before the Norman Conquest in many places. It evolved slowly over the years, but fundamentally places of worship were built close to old established rural settlements. The rapid industrialization of Britain frequently led to new centres of population far from a parish church. Mining communities, such as those strung along the Tamar Valley in Devon, for example, were poorly served by the old infrastructure.

In addition to the physical barriers of distance, many communities felt emotionally and spiritually distanced from the established church. In the 18th and 19th century the Church of England had become strongly allied to the status quo. The clergy was drawn almost exclusively from the ranks of the gentry and social hierarchy was strongly expressed within the fabric of the buildings, with strict seating arrangements based on class.

NONCONFORMIST RELIGION

Unsurprisingly, the more egalitarian attitudes of the dissenting religious groups proved attractive to many people. Some of these groups were of long standing, at least as old as the Church of England itself; others were newer. Methodism provided a new centre of nonconformist religious energy that was especially popular among industrial workers. Theology was important to many people when making spiritual decisions, but practical and social considerations also played their part. Nonconformist clergy came from much more varied backgrounds than those of the established church, many

The Methodist chapel at Morwellham, built, as it proudly announces on the front, in 1861. With a gallery on three sides, as well as seating downstairs, just over 200 people could sit comfortably and listen to the preacher.

being working people themselves who often began as lay preachers within their communities. The buildings were usually built within new communities, and paid for by public conscription, with open and egalitarian seating plans.

Calstock, just across the river was one such new industrial community; by 1902 it boasted five Methodist Chapels, two Baptist Chapels and three Bible Christian Chapels. Morwellham had two Methodist Chapels of its own while the nearest Church of England parish church was a full six miles away.

Both Methodist and Baptist ministers played a strong role in the grass-roots political movements of the day. They had strong connections with their congregations and were well used to public speaking. For many of them the duty of ministry to their flocks included care of the here and now concerns, as well as the spiritual. The history of the labour movement of the Edwardian period records the involvement of many such ministers.

Whatever denomination, the church or chapel provided a strong social centre for many people. Most had an array of clubs and activities that people could take part in. Most offered regular evenings of entertainment of one sort or another, which could be very welcome in a rural community far away from the night life of the big towns and cities. Church and chapel records list theatrical evenings, choral events, visiting musicians both classical and popular, early film shows, talks and lectures, dances and balls, even conjurers and magicians. There were girls' clubs and sports clubs, reading groups, scout groups, needlework gatherings, temperance organizations and, of course, bible classes.

THE TEMPERANCE MOVEMENT

Temperance was an important social force in Edwardian Britain. To most people temperance meant abstaining from alcohol. Many different organizations were involved with the fight against the 'evils of drink'. Some were closely bound up with the church, but there were secular groups as well. Most aimed to get people to sign up to a lifetime of abstinence and offered in return an active alcohol-free social life. The Band of Hope was perhaps the largest organization with branches just about everywhere.

Alcohol abuse was seen as one of the major causes of poverty and misery. The social norms of the time meant that it was mostly men who publically consumed alcohol, and they were depicted as depriving their families of food and indulging in domestic violence. Temperance would result in happier home lives, better-fed children, a more diligent workforce and a more moral nation. To achieve this the temperance organizations used a mix of preaching about the evils of drink with a range of rewards for those who signed up. Temperance meetings involved entertainments as well as polemics. Concerts, dances, magic lantern shows and picnics were all common. The outings were the nearest thing many people got to a holiday; they were well organized and often heavily subsidized by the wealthier members.

We create an unholy racket in attempting a rendition of a well-known methodist hymn.

SPIRITUALISM AND SEANCES

Religious opinion was deeply divided when it came to the question of spiritualism. For some people, communicating with the dead was something sinful; the spirits that were supposed to be conjured were manifestations of evil. Unsanctioned by the bible the whole practice of spiritualism was condemned as no better than witchcraft. For others, spiritualism was a confirmation of their religion, proof of the afterlife, and thus proof of the teachings of the bible. Then again, there were many who simply thought that the whole thing was bunkum, a set of elaborate hoaxes to con the credulous.

After more than a century of mesmerism, table rapping, ectoplasm, clairvoyants and spirits – and despite repeated unmasking of fraudsters – spiritualism was still very much alive and well in Edwardian Britain. Sir Arthur Conan Doyle, author of the Sherlock Holmes stories, was a fervent believer to the end of his life and there are a few references to spiritualism in his later books.

Spirit photographs were widely published for everyone to see the physical 'proof' of their existence. To modern eyes they are pathetically obvious – most being simple double exposures. But photography itself was still fairly new and carried the gloss of truth. And people were far less experienced than we are when it came to the possibility of photographic images being manipulated. Almost any accident in the dark room was happily ascribed to supernatural forces.

The Society for Psychical Research investigated a number of mediums, most notably the American Leonora Piper who claimed to have several spirit guides speaking through her. Even the mere fact that scientific investigations took place lent weight to spiritualist beliefs. At a séance the medium usually went into a trance-like state, adopted a different voice and persona, and began to give out messages for members of the audience, purporting to be from friends and family who had recently died. Some mediums preferred to work with small groups in domestic settings; others were willing to perform in front of large audiences. They were usually most successful in urban areas where audiences could expect a certain degree of anonymity. In smaller rural communities attendance at such events was often greeted with hostility from the church or chapel. It is, of course, still possible to attend such events in modern Britain.

Meanwhile, for the non believer, magic shows and entertainments tapped into the trappings of the spiritualists. A number of shows were based around devices used originally by disgraced and exposed mediums. The illusion known as Pepper's Ghost, which used an angled mirror to project an image of an actor on to the stage, had a long run of popularity. But as this became old hat, showmen moved into film. Using stop motion, stage-illusion techniques and double-exposure camera trickery similar to effects employed by the spirit photographers, ghost stories were among some of the very earliest films. *The Haunted Curiosity Shop* of 1902 made by Walter Booth and Robert Paul, for example, has a shopkeeper being plagued by the floating spirit head of an Egyptian mummy who then becomes a skeleton. The film entitled *The Mesmerist* by George Albert Smith follows quite closely the sort of experience being claimed by some practising mediums of the time.

Film ghosts were often comic figures and, in common with most of the spirit-inspired entertainments, were meant more to ridicule the gullible characters within them than to scare or convince the audiences. In fact, many of those who put together such shows were deliberately interested in exposing people who claimed to have access to the supernatural.

Anyone wishing to claim spiritualism as a serious spiritual experience found themselves increasingly held up to ridicule as the 20th century progressed.

SPIRIT GUIDES

There are fashions in everything, including the supernatural. Table rapping was definitely out of fashion by 1900, having once been the most widely practised manifestation. Hypnotism was far less of a popular draw than it had been. Ectoplasm too had fewer advocates than before. Spirit guides were, however, very much to the fore – the more exotic the better, as they had the added advantage of being photogenic. Swathes of cloth wrapped around a slightly out of focus figure, worked very nicely in the photographer's studio.

OVERLEAF Séances were often undertaken in the Edwardian period as a form of entertainment and interest rather than a meeting of believers.

CHILDREN AND SCHOOLING

 In 1889 the minimum school leaving age was increased to 12, although bright children who passed the agreed Standard IV tests could leave to take up part-time employment a little younger. (The school-leaving age wasn't raised to 14 until 1918.) Reading, writing, and arithmetic were the mainstay of the teaching, but schools were also required by law to provide needlework for the girls and drawing for the boys, as well as religion and 'object lessons' (see page 216). History, geography and physical education were also taught in most schools although they were not compulsory.

THE SCHOOL HOUSE

The school house at Morwellham was a single room, as so many were. It was licensed for 60 pupils, which would have been a most terrible squeeze. But by the end of 1902 there were only 26 pupils attending regularly, as so many families had moved away when the Devon Great Consols Mine went into receivership. Twenty-six pupils to one teacher was an extremely good ratio for the Edwardian period when classes of 70 or 80 were commonplace. All the children were taught together, regardless of age.

Children were often packed in like sardines in rural schools.

The schools of Calstock and Gunnislake across the river were much more representative of Edwardian schooling. Calstock had two schools in 1902; an infants Board School whose one qualified teacher, Miss Eliza Moore was in charge of 114 children, and a National School taught by a husband and wife team, Mr and Mrs Cann, where 118 children regularly attended. Along the river at Gunnislake, in the same year Miss Alice Smith looked after 150 infants at the Board School, with Mr Robert Leverton

and Miss Olive Whitford teaching 163 of the older children. The other Board School in the town had a staggering 190 pupils under the professional care of Mr John Morgan. Board Schools were so called because they were overseen by a board of governors as specified under the 1870 Elementary Education Act – as opposed to church schools, which were, unsurprisingly, run by the church.

The professional teachers had some assistance, although not from qualified staff. Pupil teachers were an important element of the Edwardian education system. These were bright pupils who stayed on at school beyond the usual leaving age and assisted the teachers in exchange for further tuition themselves. For the teachers of such enormous classes they represented the only possible way of coping. Once the school day was over, an hour or two was devoted to guiding the studies of these youngsters, a few of whom could go on to some kind of secondary education or even teacher training college. It wasn't an ideal system, but it was cheap and offered a working-class child, male or, even more radically, female, an opportunity to change their life chances utterly.

There was a move during the Edwardian years to do away with the pupil teacher system entirely. Schools wanted teachers who had attended a high school and a training college and passed a teaching qualification. Training facilities for teachers expanded rapidly in the first few years of the 20th century. Many of the older established colleges were funded and run by the churches of various denominations, but there were also an increasing number of secular establishments and by 1910 a full 16 university colleges had also established departments specifically for training teachers. The nearest to us in the Tamar valley was St Luke's in Exeter.

Ways of Learning

Much of the curriculum, especially for the younger children consisted of rote learning and drills of one form or another. This included physical training, which was either derived from calisthenics or from army drill books. Devon schools were noted for the latter, with several schools employing veterans to lead the children through patriotic exercises with broom handle guns.

Some subjects allowed for a little more imaginative teaching and learning. 'Object lessons' were purposely designed for this, and the teacher was encouraged to use real things or pictures to teach a range of subjects – from some sort of nature table to how sailing ships work. At secondary school level, science lessons usually gave most scope for imaginative and exploratory learning.

With pupil teachers being eased out of the system, one form of opportunity for working-class children was being lost. At the same time, another began to open a little wider. In 1907 an act came into force that extended the provision of scholarships and free places in high schools. All grant-aided secondary schools were now required to offer at least a quarter of their places free of charge to bright local children. In 1904 the Board of Education had set the curriculum for these such schools and meant that children were now able to study for a further four years. It included English language and literature, geography, history, a foreign language, maths, science, drawing, manual work, physical training and, for the girls, household crafts. It's a curriculum familiar to most of us in Britain today.

I own several Edwardian school textbooks, which give a flavour of what children were taught. The geography textbook belonged to my great-grandfather, who benefitted from one of those free places at secondary school. It is a book of lists to be learned by rote. Each country of the world is described in terms of its principal towns, exports, rivers and mountains. Political geography is represented by simply noting where a country belongs to one of the great empires. A short section on the rotation of the earth and its various climates obviously held a little more appeal for my great-grandfather for he has added a few notes of his own, working out things like miles per second in the margins. Longman's *Complete Course of Needlework, Knitting and Cutting Out* by Miss T.M. James was a textbook for the teacher rather than for the pupil to use. Needlework classes lasted a minimum of two hours a week for girls, while the boys continued with the three R's.

Miss James begins her complete course with thimble drill. This is aimed at three- and four-year-old girls, who would repeat the actions every day for several weeks, adding a number of other drills to their repertoire as they went along – including needle drill, position of work drill and stitch drill. Later they were shown large-scale demonstration frames of how to do the stitches and the little girls began to do their first hems and seams. By the time they left school at the age of 12, girls were expected to be able to make the family's underwear, nightclothes and aprons, to have a repertoire of some thirty stitches, to be able to darn and mend, and also undertake simple knitting projects.

I have a 1903 science textbook for high school girls: *The Science of Home Life, a Textbook of Domestic Economy* by W. Jerome Harrison. At first glance I was somewhat taken aback that the science education of girls was so thoroughly bounded by their domestic role. The biology and chemistry, however, is well presented, with a range of experiments concerned with the nature of oxygen, water, electricity, foodstuffs and cleaning compounds. The girls were obviously expected to learn to use microscopes and be able to carry out a range of chemical tests. For example:

What water is made of? – Let us take an electrical battery, consisting of several cells joined together; this battery will produce a powerful current of electricity. To each end of the battery a wire is fastened, and to the other end of each wire a thin strip of the metal called platinum is attached.

Remember this text book was written in 1903, well before the national grid existed. For children in rural areas gaining a secondary school education could be difficult. Even if free places were available to bright pupils, transport to school could be difficult if not impossible. The majority of Edwardian country children were out working on the land full time by the time they were twelve.

By the age of twelve, most rural children were out working full time. The lads who came to join us for the day relished being put to work, despite the pouring rain.

5

LOCAL INDUSTRY

MARKET GARDENING

The Tamar Valley sides are steep, but the soil is good and fertile and the climate exceptionally mild. In the early years of the 20th century market gardening was a mainstay of the local economy here. Fruit and flowers grew well in the soil and the warmth and sun meant that crops could be harvested very early in the season when London prices were high. As you walk about the valley it is possible to see between the trees the remnant of walls and terraces that formed the market gardens almost everywhere you look. Forestry Commission plantations cover much of this ground now; other terraces have disappeared under natural regrowth. Just above the canal is a small shed-like building now surrounded by plantation pines. It was once the packing shed for two large flattish gardens of daffodils, a small outside loo straddled the stream for the benefit of those working in the gardens.

For a good two weeks the hoe and fork were Alex's only company on the market garden terraces.

CLEARING THE GARDENS

A casual glance at the heavily wooded hillsides now makes it hard to imagine the cleared, terraced valley slopes of the Edwardian years. Whole hillsides were planted with large widely spaced cherry trees, with the ground beneath growing acres of daffodils: the smell of strawberries hanging in the air in the early summer.

Some of the gardens, such as those up at 'peace' occupied the more gently sloping ground, but in other places the terraces are steep and small, clinging to the hillside to catch the best of the spring sunshine.

Thankfully one area of terrace had been cleared a few years ago and we were able to take this on. It is not possible to get a horse and plough on to such small scraps of land

or such steep slopes. The area is about the size of six standard allotments, and it must be dug, manure, planted, weeded and harvested all by hand. It is only the great productivity of the site that made this a commercial prospect in the Edwardian period.

Although we had no trees growing on our market garden terraces, there was still a lot of work to be done. The ground needed clearing from last year's vegetation, and the walls and steps all needed a little attention to remove the brambles and ivy that had begun to take hold.

We began by enlisting the help of our goats. Goats notoriously will eat anything while their light frame means that they do very little harm to walls. If, on the other hand, you have an area of forest floor to clear, then pigs are the perfect choice since they will pull up and eat everything except the trees, leaving the ground clear, turned over and fertilized. But in this situation pigs would damage our precious terracing, rooting out the stones to get at the bramble roots between them.

We tethered the goats on the garden and left them to get on with it. As is the way of goats they preferred to ignore anything beneath their feet, keeping their attention on the plants growing in among the stones of the walls and steps. They did a fantastic job,

We needed to clear the garden slopes of last year's radish crop before planting out our strawberry plants.

RIGHT *It wasn't long before the old greenhouse was full of young strawberry plants.* OPPOSITE *We stand with one of our finest gardening assistants – a goat.*

cleaning the whole area thoroughly and without damage. Once one area was bare we moved them along to the next patch.

We soon found it is easy to control a goat that is happily eating its favourite food, but nigh impossible once that food source is exhausted. Better to keep an eye out and move them on to a new patch pronto rather than forget about them and spend the next half hour chasing a goat that has pulled up its tether.

We all had a go clearing the ground whenever any of us found a spare moment, but it was hard going. I found a mattock the easiest method, but Peter favoured a spade. Alex tried just about every tool going but usually ended up with one of the hoes. We were hoping to get the soil clear before any of the many weeds had a chance to set seed. A forlorn hope, time was catching up with us.

As we worked we came across a couple of patches of strawberries which had been planted by our predecessors on this patch of land. We cleared around them and carefully pegged out the new runners into tiny flower pots to be overwintered in the old greenhouse at the top of the slope. To start them off we left them attached to the parent plant and buried the pots in situ. This gave the runners a chance to develop roots before we severed their connection with the parent plant and let them grow on, on their own. All in all, we were able to gain around 200 new little plants this way for no outlay.

PETER'S DIARY

One thing that went into market gardens in abundance was work so we made sure the patch that we took on was small but manageable. We were lucky enough to get advice from a number of locals such as Stan Sherrell who had been a market gardener back in the day. One thing that he drummed into us was that the soil had to go uphill: any digging or hoeing always had to be done up the slope.

Stan also gave me good advice about planting gooseberries. They needed to be two and a half feet (76cm) apart, with the rootball or 'heart' below the surface, with all the roots facing down. Any stems below the heart could be removed. His advice certainly worked as the bushes that I planted are thriving.

THE STRAWBERRY STORY

The market gardens of the Tamar Valley began to take on a new importance in the local economy in the second half of the 19th century. Before that fruit and vegetables grown on the sheltered slopes had fed the local population, the produce selling in markets along the valley and down the river to Plymouth. The railways began to offer new opportunities to the more enterprising grower. Chief among these was a certain James Walter Lawry.

In 1862 22-year-old Lawry went up to London to visit the Great Exhibition at Crystal Palace.

'After seeing the show, having heard of Covent Garden Market, my friend and self determined that we would rise early and visit this renowned market whilst yet business was in progress. It was early in June and to my surprise, I found that there were no out-of-door grown strawberries offered, whilst at home the crop was nearly finished before we left. On enquiring the price of the hot house fruit offered, I was staggered at the difference from that we had been receiving at Devonport.

'I got into conversation with a salesman named Israel, and explained that I hailed from Cornwall and was a grower of strawberries which were now practically finished, although if he would undertake the sale I would write the people at home and get them to forward a small quantity as an experiment. He promised to do his best, and I made the venture…'

Over the next few years Lawry both quadrupled his own acreage and acted as an agent for his neighbours. The London price held at roughly four times the local price for best-quality early fruit, making it well worth the additional transport costs.

Taking strawberry runners to form next year's new plants. Our strawberry crop in its chip baskets all ready for dispatch to London.

LEAFLETS FROM THE BOARD OF AGRICULTURE AND FISHERIES

 In 1904 the Board of Agriculture and Fisheries decided to collect all its published advice to farmers together in two volumes, with a third volume joining them in 1915.

This was officially sanctioned government advice, drawing on the work of botanists, chemists, lawyers, vets, naturalists and farmers, and represented what was thought of as best practice for its day. The leaflets give us a snapshot of the new direction that the Edwardian countryside was moving towards.

In addition to scientific knowledge, the leaflets tried to offer commercial advice, laying out the likely costs and returns available from a variety of different types of farming – from strawberry growing to making soft cheeses. Others in the series offered legal advice, outlining the new responsibilities and rights of farmers.

The leaflets have proved enormously useful to us in our farming endeavours. Leaflet no. 207 Strawberry Cultivation is a mine of information. First published in 1910, it gives practical advice on manuring, planting times and spacings, appropriate varieties, cultivation techniques, pests, diseases, and how to control them. It also discusses packing options, labour costs and current prices both for manures, plants and yields. The advice on growing daffodils – Leaflet no. 224 from 1909 – was obviously well founded, as Tamar valley growers were still using the same techniques within living memory.

New thinking about rural life is everywhere in the leaflets. For example, some of the early leaflets concern the natural history of birds. There was a long tradition of trapping and killing birds that either damaged crops or took young game birds as prey. The Board of Agriculture's concern was to stop the trapping and shooting of those birds which – rather than injure a farmer's livelihood were in fact helpful in controlling pests. Information on barn owls included analysis of their pellets, showing how many mice they consumed: 'in 700 pellets of this owl there were found the remains of sixteen bats, 2,513 mice, one mole, and 33 birds of which nineteen were sparrows'. The leaflets encouraged the use of nest boxes to increase the numbers of useful insect-eating birds such as starlings, flycatchers and swallows.

There were some aspects of government advice, however, that we had no intention of following ourselves. The Board of Agriculture was at the forefront of the new chemical attack on pests and diseases, and its leaflets encourage the use of a huge variety of pesticides and fungicides. Many of these chemicals are now banned for the harm they do, both to the environment and to the people who come into contact with them. Leaflet no. 69 recommends a spray of arsenate of lead for use on fruit trees. As both arsenic and lead are extremely poisonous we had no desire to try this out – nor would it have been legal to do so nowadays.

By the Edwardian period it had become a well-established trade. Several factories were established producing punnets and baskets for packing the fruit, which now included apples, plums, raspberries, rhubarb, gooseberries and cherries. The railways too had responded to the business. In 1890 the London and South-Western Railway extended its line out to Bere Alston and Bere Ferrers to cash in on the volume of fruit traffic in the area. In 1908 the newly completed Calstock Viaduct added further convenient collection points.

The Board of Agriculture and Fisheries advice on packing and grading strawberries would have come as no surprise to the people of the valley; they had been doing it for two generations.

'Fruits such as strawberries and raspberries are worthy of much care, strawberries being sorted into at least two grades and sometimes into more. The best are placed in punnets, the next in small boxes and a third grade can be sold in boxes or baskets holding from 6lb to 12lb...

RUTH'S DIARY

The strawberry plants arrived at last today. We had ordered 300 plants. Being so late in the season we had asked for quite mature plants in the hope that we could catch up and get a decent crop in the spring.

Luckily for us several of our neighbours agreed to give us a hand. Many of them had been in the strawberry-growing business themselves and were able to stop us from making too many mistakes.

The large plants that we had thought would give us a head start were looked on askance. 'You'll not get much fruit off them' was the consensus of opinion. 'Small plants have more vigour, see'. Thankfully the large plants came with runners and small plants attached so it would just be a matter of sorting and preparing the plants. This proved to be my job.

It was one of those raw bitter days although the sun did make a few appearances. Alex marked out the rows and I stood at the top of the bank working my way through the tangle of plants, sorting out the plants into sizes, pulling off all the dead or overblown foliage and trimming the roots. Once Alex got started with the planting it was a struggle to keep up with him. The soil on the plants was close to freezing point and the cold wind made it feel all the colder. Each

RUTH'S DIARY

plant had to be carefully picked over: the roots straightened out and trimmed to the right length so that when Alex dropped them into the hole the roots would all remain pointing down and none be bent back up. Peter was doing a final clearance on the next terrace. As the day wore on our neighbours left us to it, confident that we had finally got the hang of strawberry planting.

Not much banter today, we were all too cold for that. I had the easy job without a doubt, but I was still very glad when it was time to dash off and put the dinner on.

'Strawberries may be packed in lots from 3lb to 6lb of selected fruits, but the first named quantity is best for the finest fruits and the smallest railway boxes just hold that amount conveniently, allowing for a little packing material at the top and bottom...

'The finest early strawberries should be packed in 1lb punnets, which may be either deep or shallow, round plaited chip punnets, or square ones (with or without handles). The round punnets are best packed in trays with lids, and those generally employed will take six punnets. They are only used for the earliest and choicest of fruits, when prices are good...

'The question of branding or labelling must be carefully considered... growers who intend to make a substantial business, and who deal honestly in the best produce, should have their own names on the packages. This is sometimes objected to in a market, but if a grower cannot make his business through the ordinary channels he must try fresh ones. It is best to endeavour to supply the shopkeepers, or to develop a trade with private customers, and send direct to them.'

MAKING STRAWBERRY JAM

While the best quality early strawberries could be fetching up to £40 an acre the late season fruit could fetch as little as £10 an acre. These fruits were largely destined for the jam factories. Many a local family went on to jam production themselves, sure of a good market supplying jam for the cream teas of the summer season. The best jam recipes are always the simplest.

INGREDIENTS
2 lb (1 kg) sugar
Just enough water to wet the sugar
2 lb (1 kg) fruit

METHOD
Put the sugar in the pan and begin to add a little water. Put the pan on the heat and stir, adding a little more water until all the sugar is just barely dissolved. Bring the syrup up to a rapid boil and then throw in the fruit. Stir very gently so as not to break the fruit. Boil until the jam begins to set or until it reaches 219°F (104°C) if you have a sugar thermometer. Pot up immediately. Strawberry jam is better for being only slightly set and rather runny as over boiling spoils the flavour.

Strawberries were picked in the early hours of the morning to keep them cool for their journey to London's Covent Garden Market where they could be sold at astronomical prices.

MARKET GARDEN PESTS

Spring was upon us and everything was growing wonderfully. The oats were rocketing up in the warm moist weather, the potatoes had emerged, lambs and calves were skipping around the fields and chicks were hatching. The problem was, every other creature and plant in the valley was also growing at a rate of knots and many of these were a potential hazard to our farming and market gardening ambitions.

Insects would plague the livestock – particularly the sheep – and weeds would threaten the potatoes, but the most vulnerable of all our crops were the strawberries. The gardens where we had planted them were terraced with stones and, beautiful as they were, the dry-stone walls were a haven for slugs and snails. These invertebrates attack all crops: in a case reported from Yorkshire in 1909, they were responsible for the destruction of over half the crop of oats and barley while at the same time seriously damaging the grass fields.

Organic gardeners sometimes recommend laying a few stones to shelter slugs. Each day you turn over the stones and destroy the slugs underneath. Slowly this method can help to reduce their numbers. For us, however, it was useless. There were hundreds, if not thousands, of stones embedded in the terrace walls, concealing a whole army of hungry slugs. I decided on a three-pronged attack. First I looked up what the Board of Agriculture and Fisheries had to say. Leaflet no. 132, Slugs and Snails, described the many varieties of these pests. I was horrified to read of one particular type, the 'Strawberry Snail', which is 'particularly troublesome in strawberry beds'. Having spent a good hour or so collecting snails and slugs, I was mortified to find a significant number of them in my bucket. The leaflet recommended using an irritant dressing to protect plants. This could be soot and lime, salt and lime, powdered coke (a form of coal) or – most potent – lime and caustic soda. In my state of panic I decided to go for the latter and mixed up a dressing of 1 part caustic soda to 24 parts lime powder. I laid this in a thin trail around the various strawberry beds to create a barrier no slug or snail would be able to cross. My good friends the blackbird and thrush would hopefully deal with those molluscs already in the bed and I'd remove as many as I could by hand, but those outside my ring would never get in.

One of the many things slugs have in common with Peter is their fondness for beer. My second weapon was going to be the good old-fashioned beer trap. This is a jar half full of beer, set into the ground and partially covered with a stone. I put a number of these at strategic locations, but I didn't want to overdo it and have Peter accuse me of wasting beer. Finally, my *pièce de résistance* was the ultimate combatant – the duck. I drafted a couple of the greedier-looking Aylesbury drakes down to my market-garden war zone. Ducks have to be the most gluttonous of all farmyard animals. Within seconds of releasing them, they were ransacking the bracken mulch in search of slugs, snails and any other grubs they could lay their beaks on.

THE CHERRY HARVEST

In our market garden we had concentrated on daffodils, strawberries, gooseberries and redcurrants. However, one of the main crops produced by the hard-working women and men of the valley were cherries. Cherries were so abundant in the valley that it has often been compared to Japan and a cherry feast was held each year (up until 1937) at Pentillie Castle. At Morwellham Quay when walking down the side of Barn Field you can see the rows of cherry trees that once cropped abundantly; like the daffodils planted below them, they still produce a crop but not one that is large or easily accessible.

To experience a cherry harvest we headed over to an orchard on the Cotehele estate, where we made our cider. Here we met up with Jess Collins, who has recently taken charge of the orchard and is bringing it back to life since it was last harvested more than 20 years ago. We were also introduced to Norma Chapman, whose family were market gardeners and who has worked with her father tending and picking their cherry trees. Both ladies were very knowledgeable on the subject of cherry growing and they gave us a good idea of this aspect of market gardening in the valley.

As with any crop, the main concern is keeping enough of the fruit on the tree to make a profit and that means deterring the predators: in this case birds – starlings can clear a tree overnight. We had heard that nets were often employed to cover the cherry trees (which does make sense in an area that has an abundance of fishing nets), but Norma told us that her family used to tie ropes between the trees and hang metal objects from them to make a noise.

Assuming you have managed to protect your crop, then it's harvest time. To do this you need ladders. Ladders were housed in a purpose-built building that kept them under cover but allowed the air to circulate, maintaining them through the winter. The largest ladder in the ladder house was 41 bars and the shortest was around 20. The trees where pruned specifically for easy ladder access: so if a stout limb protruded from the trunk, it would be left as a ladder rest. When picking we soon learned that moving the ladder took time, while a well-placed ladder enabled you to reach lots of cherries.

Norma stressed that safety was paramount when it came to going up the ladders: if you are injured you can't pick, and if you can't pick you won't make money. Considering that the 41 runger didn't even reach the tops of their cherry trees, coupled with the increase in perception of depth that being up a tree on a slope gives you, I can see why. The ladders were tied on to the tree using a rope attached to a loop that went around the ladder; then it could be easily slid up and down when repositioning, saving valuable time. The feet of the ladder were dug into the ground, especially when working down slope from the tree.

Once we had our ladders in place Alex and I were up the tree picking. We were advised to start at the bottom and pick everything in reach and then move up, repeating

A pannier full of freshly harvested cherries ready for Laddie to take back to the farm.

the process. What we picked we placed into baskets – bucket shaped and made of cane – which were hung from the ladder using hooks and rope, so that they were easily movable. We had to pick the whole cherry, including the stalk, otherwise all the juice would run out and the fruit would be unsaleable. On the day we went to the orchard to harvest our cherries it was stunning sunshine; Norma told us that when they picked in the rain they would lay the cherries out on sacking undercover to dry, otherwise they might split.

A cherry ladder, like a thatching ladder, has vertical stiles that are curved on the rung side, as the picker will more than likely be leaning over and this means they will not have to lean against a hard straight edge.

The harvest was usually just a family affair but occasionally Norma's father would get in help. In their worst year they got only 20 lbs (9kg) of cherries but in their best year they had over 8 tonnes . On average they collected 3–4 tonnes. On hearing this, Alex and I doubled our picking efforts. We loved doing the job and I can see why Norma says that her family would 'sing their hearts out up the trees'. She also said that it was lovely to hear their voices echo right out across the valley.

In the orchard there were many varieties of cherry tree such as Bullions and Birchenhayes but we were picking the local Burcombes, which are a large black good mid-season cherry with a shorter stem. Jess indicated the bulge on the Burcombe trees that shows where they had been grafted as young plants. Norma can remember her father grafting a cherry tree and by the time he retired they were picking fruit from it using a 20 bar ladder.

The cherries were preserved in Kilner jars to ensure cherries all year round in the Chapman household. Both Jess and Norma said that when picking they don't eat the cherries. Firstly it is counterproductive and secondly there are only so many cherries you can eat. But I think I managed to eat as many as I picked – and I picked a lot.

PICKING DAFFODILS

The slopes of the Tamar valley had been gardened very intensively from the 1860s and by the Edwardian period in the 1900s this had begun to cause some problems of its own. Intensive agriculture of any kind in time leads to a major increase in pests and disease. Forty years of major strawberry growing had led to a serious build up in a number of strawberry-specific diseases, which were cutting the yields for many growers. The Board of Agriculture and Fisheries recommended a whole host of chemicals to help the grower combat these problems but for many growers it made much more sense to switch crops.

Daffodils require very similar climatic conditions to strawberries, and the early blooms grown on south-facing slopes offered similarly high prices at market.

The hand cultivation, which local growers were so adept at, was easily transferred to daffodil growing.

The beginning of the 20th century saw the Tamar Valley slopes begin to change colour as grower after grower tried his hand at flowers, and especially daffodils. In the very early spring it was awash with yellow, then later the local speciality Double White daffodil took over, just as the cherry blossom began to bloom overhead.

Packing the Daffodils

'Market prices vary considerably… 9d to 1s per dozen bunches may be taken as an average, although the small cultivator will find that where special attention is paid to careful picking, packing and grading, rather higher prices will be obtained.'

Board of Agriculture and Fisheries Leaflet no. 224, 1909

Marion and Iris at Denham Farm kindly showed me how they used to bunch and pack the daffodils for market. Green tissue paper for yellow daffs and blue paper later in the season for white ones.

The little market cart was ideal for carrying the flowers off the steep slopes to the local station.

Each carefully arranged and tied bunch had to be laid in the box in strict sequence.

The packing shed bordered the gardens and large windows looked out over the valley. A chimney in the corner allowed a small fire to keep the worst of the chill off and help dry out the blooms before boxing them up. Trestle tables lined the walls and bucket after bucket of daffs filled every spare bit of floor space. Raffia, split to make it go further, and cut carefully to length was hooked on a nail near at hand. Standing at the trestles we sorted daffs into bunches, twelve to a bunch arranged in four rows of three, tied around with a wisp of raffia. The bunched flowers were returned to buckets of water to stand overnight and firm up. Yesterday's bunches were packed carefully into the boxes with all the blooms facing up, the tissue folded around and the lids popped on before they were loaded into the cart to begin their journey to market.

I found it amazing just how closely Marion's and Iris's memories of packing daffodils in their childhoods matched the official advice given by the Edwardian Board of Agriculture and Fisheries: *'Packing... all boxes should be papered, allowing sufficient paper to project over the sides and ends to cover the blooms completely when packed. The boxes should not be unduly crowded, as a lesser number of bunches nicely tied and carefully packed will realise more than an increased number of bunches crowded into a similar box...'*

Most of the Tamar Valley daffodils went up on the train to Covent Garden flower market. Each of the local stations took on additional staff during the season, who not only loaded the goods but telegraphed ahead to all the receiving points on the journey to arrange handling at the other end and alert the wholesalers. Until the Calstock viaduct was built to carry the railway over the river, growers on the opposite side of the valley delivered their flowers to the quay at Calstock where the railway had a boat and staff waiting to take them across to the station.

As well as selling at Covent Garden, some growers sold their flowers direct to florists and hotels and many others sold through the local markets. The most important of these was Devonport in Plymouth. On market day several large paddle steamers plied the Tamar, picking up cargo from a series of quays along the river. The river was regularly dredged in Edwardian times, making it possible for large boats to travel right up to the weir below Gunnislake, at any state of the tide.

THE COOPERATIVE MOVEMENT

 'The establishment of these societies in the rural villages in which they are found has evidently not only added to the prosperity of many of the villagers, but has stimulated neighbourly feeling by showing men how they can help their fellows by the exercise of care and mutual trust, without any real pecuniary risk to themselves, has encouraged thrift and efficient methods of cultivation, and has at the same time increased self respect of the individual members, and inspired them with hopes of progress.'

Board of Agriculture and Fisheries Leaflet no. 260

This is perhaps a rather strange sentence to find in a government publication, yet it refers to setting up agricultural cooperative credit societies, and reads frankly more like an article of religious faith than anything else. Cooperatives of one form or another formed the great hope of the Edwardian era. With trades unions and the labour movement becoming important movers and shakers in the political world, cooperatives were seen as a bridge between people's aspirations for a fairer society and a continuation of the traditional values of capital and self-help.

In a cooperative, the business is owned collectively by all members who work together for mutual benefit. The Board of Agriculture and Fisheries recommended several different types to the farmers of Britain, including dairy farmers' cooperatives, whose purpose was to ensure a better price for milk and market-gardening cooperatives to secure better transport costs. Friendly societies – a sort of early insurance scheme – to ensure livestock against sickness were described and advocated in Leaflet no. 221. Cooperative mills to process produce were another recommendation, as were factories to manufacture packing materials for agricultural uses, and cooperative banks to look after their finances.

Edwin Pratt in his 1909 book *Small Holders – What They Must do to Succeed*, sings from the very same hymn sheet. 'They should group themselves into registered cooperative societies… and then supplement their land society by a variety of other cooperative arrangements, including therein the joint marketing of their produce, or, at least, its joint transport, by motor wagon or otherwise, to the nearest town or the nearest railway station.'

The aim was to help keep the small agricultural producer in business, to bolster the numbers of small-scale capitalists, to provide a way of levelling the playing field without overturning the status quo. Cooperatives operated in many other walks of Edwardian life, from funeral provision to local shops. The market gardeners of the Tamar valley did eventually form a cooperative but not until 1921 when it was founded by a local schoolteacher with a motto of: 'One for all and all for one'.

But while there was a positive rash of cooperatives set up in the Edwardian era, not many of them were long lasting. It proved difficult to come up with a formula that served everyone's needs. Many a society failed when one bad apple in the membership barrel got greedy. The few cooperatives that have a truly long-term history are those that got their internal governance right.

THE HATCHERY

Mining may have enriched the Bedford estate but it killed off a large proportion of the fish stocks in the river Tamar. It was said that if you placed an iron horse shoe into the Tamar during the height of the mining industry it would come out copper plated. When mining became unprofitable the Bedford estate turned to other sources of income, one of which was high-end tourism. During the Edwardian period the Duke of Bedford took on the task of turning over a large part of his concern at Endsleigh House to fishing, modelling it on Scottish retreats. He planted Douglas firs, stocked up on deer, changed the title of groundskeeper to ghillie and secured the fishing rights for a 14 mile (22km) stretch of the river from Greystone Bridge to Gunnislake. All he needed now were the fish for his rich London clientele to catch.

Judging by the confused look it seems they don't call it a baffle box for nothing.

Fish farming has a long history but the methodology developed in Edwardian Britain was, to an extent, unique. The Duke of Bedford decided to set up a hatchery at Endsleigh with the sole purpose of replenishing the fish stocks in the river. To do this he employed the services of Mr McNicol in 1901, who had previous experience working in Scotland where this technology was being pioneered. The Endsleigh hatchery was a thatched hut where the eggs were hatched out before being transferred into little ponds and brought on to different sizes before being released into the river. The fish that were being cultivated were salmon and trout. The varieties of trout not only included the well-known brown and rainbow but also the more exotic sounding golden pink-eyed.

As we are looking at mining in this project, because it has had such an impact on the landscape and the socio-economic climate of this area, I thought it only fitting that we should also explore the practices of restoration executed in the Tamar Valley. And building a fish hatchery would be a fitting project. I called upon the services of Horace

Adams and his son-in-law Rodney. Horace's father had worked at the Endsleigh hatchery; he came home one day in the summer of 1933 to say that Mr McNicol had proposed talking to Horace about a job. Later that year Horace (who had never even held a fishing rod) found himself working at the hatchery too and has never looked back. He went on to write the book *Endsleigh: The Memoirs of a Riverkeeper.*

Horace and Rodney talked me through the nuances of the science and the art of raising fish; what was likely to go wrong and the problems that I would come up against. Horace also told me of the age of 'trap and shoot': a crueller, simpler, but often happier time. It was not only the responsibility of those working at the hatchery to breed the fish but also to protect them. Their enemies included the kingfisher, the heron, the cormorant and the otter – all of whom would have to be dispatched by one means or another if the opportunity arose.

THE WATER SYSTEM

Surveying my hatchery; now all I need are the fertilized trout eggs.

The first step was to find a suitable spot for my hatchery. I needed running stream water for a constant supply of oxygen to the eggs and later the fish. My first thoughts had been on the quayside, but it is prone to flooding and that would be disastrous as the river is brackish at Morwellham. In its heyday the quay was largely powered by water. Various waterwheels drove the incline railway used by the mines; waterwheels situated down mines ironically pumped out water. Waterwheels supplied drinking water to the village; powered hammers to pound up ore; thrashed the farmers cereal crops; the list goes on. Much of this infrastructure is now defunct and disappeared along with the miners but in its place a hydro-electric station was built in the 1930s, which to this day utilizes the power of the water in the valley.

However, there are still a few streams that flow down the hillside and Anthony Power (a former manager of the site) took me on a guided tour of the water system, starting at the entrance to the Tavistock canal tunnel that is 1¾ miles (2.8km) in length! The canal on the Morwellham side of the tunnel is a contour canal that follows the hillside, ending at Canal Cottage, just above the farm. This canal was used by the market gardeners to transport their produce, but it is now dry as the water flows to the reservoir of the power station instead. I picked the strongest of the streams that still flow to the mill pond near the farm because I didn't want it to dry up or freeze over. (A good decision in hindsight as the other stream proved to be somewhat ephemeral.) Plus the stream I chose had cut deep channels in the ground with mini-waterfalls, resulting in height variation. This was ideal for me as my hatchery box would have to be lower than the incoming flow of water.

The mill pond would also make an ideal release site for the fish when they hatched. The pond had acted like a header tank to power the large waterwheel that still existed in Morwellham Quay and had recently been restored by Simon Summers, the master blacksmith. In its time this wheel was used to pound up manganese ore, but now it just served as a reminder of Morwellham's industrial heritage. There is also a ram pump just below the mill pond which pumps water up to another header tank for the waterwheel at the farm. Originally the farm was fed directly by a leat that runs from the canal; the ram pump is a Victorian installation that works on a pneumatic principle. The energy of 100 pints (57 litres) flowing through the ram pump forces 1 pint (0.6 litres) up the hill.

BUILDING THE BOXES

Having found an appropriate site for the hatchery my next step was to design and build two boxes: a baffle box to regulate the flow of the water and to filter out the majority of the silt; and an egg box that would contain the fertilized trout eggs and provide an environment in which the newborn fish could live, assuming that the eggs survived the winter. The baffle box was 1ft x 1ft x 3ft (30cm x 30cm x 90cm) with an inlet for the

Persuading the torrent of water to flow into the baffle box was no easy feat.

water and three vertical boards to disrupt the flow. It was connected to the egg box via a cast-iron drainpipe. I used a cast-iron pipe to connect the two boxes as I knew the water may well contain high levels of copper and this would filter some out.

The egg box had two initial compartments designed to act as additional baffles, then three sections to house the eggs. I lined two of the sections with stones from the river bed to create a natural habitat for the eggs. For the third section I made a tray from pairs of glass rods to hold the eggs. Glass-rod trays were in use in the Endsleigh hatchery during the Edwardian period and enable the fish farmer to check the eggs and remove any bad ones. The final section of the egg box was an open area into which the fish could venture once they reached the fry stage and started to look for food.

By the start of the 20th century steam-powered saw benches were commonly in operation and regular-sawn timber planks were much more available. I built the two boxes out of pine planks, cutting them to length using an ornate carpentry saw dating from the mid-19th century. The key is to use wood that hasn't been treated in any way so that it doesn't harm the fish. To fasten the planks together to form the relevant boxes I used screws. Screws started to become a common form of fastening at the end of the 1700s when Yorkshire-born Jesse Ramsden invented a lathe for cutting the screw. Standard screw sizes were initially introduced to Britain by Sir Joseph Whitworth in the 1840s, and the British Standard Fine thread was established in 1908.

Although the joints in the boxes were tight and the wood would swell when soaked in water, I needed to ensure that there were no leaks. To do this I painted the inside of the box with pitch. Pitch is an amazing substance. Today it is largely derived from petroleum and is known as bitumen, but historically pitch is a resin extracted from wood. Despite its solid appearance and tendency to shatter when struck, pitch is actually a viscoelastic polymer, which means that it is in fact fluid and over time it will flow. Pitch drop experiments have shown that it takes about a decade for a single drop of pitch to fall out of a funnel full of the substance. To make the pitch pliable enough to paint on to and into the joints of my two boxes, I heated lumps of it in a pan over a cast-iron stove.

Once I had made and sealed my baffle box and my egg box, it was time to take my hatchery to the chosen site. I placed the baffle box in the stream at the base of one of the level drops in the water course and used a couple of short pieces of drainpipe to direct the water into it. I put the egg box on the bank of the stream (lower than the outflow of the baffle box), balanced on flat tiles so that it was both solid and level. Now all I needed was my trout eggs.

STOCKING THE HATCHERY

The hatchery had been set up a few days earlier, to make sure that it was fully operational and to iron out any teething troubles. I put some eggs on the glass rods in the first compartment. The remaining eggs I divided between the other two stone-lined compartments. The key to survival was going to be protecting the eggs from sunlight, making sure they didn't receive any nasty jolts and ensuring a flow of water that would replenish and maintain the oxygen levels in the hatchery. To do this I put a lid over the egg chambers, I built an external hoarding around the egg box to keep out any animals and also protect against the harsh winter (I also packed straw between the hatchery box and the external hoarding) and I covered the baffle box and the ends of the inlet pipes with chicken wire to exclude the constant influx of leaves that came down stream.

I checked the hatchery twice a day for the first few weeks and then reduced this to once a day when I was happy that everything was going smoothly. My major battle was against the leaves, which were a persistent problem until after January. I also had to carefully lift the lid and remove any bad eggs. A bad egg goes an opaque white colour, which means that it has spores on it that will spread to adjacent eggs unless removed. I did this by very carefully using a pair of tweezers that I had fashioned from some wire on Horace's advice (and later using two twigs like chopsticks when I mislaid my tweezers).

The initial stage in the hatching process is 'eyeing up'. At this point the eggs develop a pigmented black eye, which is the embryo of the trout. The eggs also become less

STRIPPING THE EGGS

Being an Edwardian farmer for a year means undertaking some very unusual tasks, but this has to rank as one of the strangest. Helped by Trevor Whyatt, a trout farm expert, who had caught several female trout that had come up river to spawn, we stripped the fish of their eggs. This involves picking the fish up with one hand and with the other gently but firmly stroking its underside, so that the eggs come popping out in a continuous flow (which we caught in a bowl), until you can feel that the egg sac is empty. The amount of eggs that each fish produces is related to their body weight; for each pound (½kg), a female will produce between 600 and 800 eggs. Initially the eggs are very sticky because they are soft; they need to be mixed with water to harden them. This is also the stage to introduce the milt – the seminal fluid of the male trout. Very little milt is needed to fertilize a few thousand eggs.

In nature the female swims upstream to a suitable site, makes a hole in the gravel with her tail and lays her eggs (often older female fish that are caught in rivers have tails that are virtually worn away from digging in the stream bed). The male trout then swims to the area and releases his milt. The female then covers the eggs up again and leaves them to fend for themselves. Needless to say, the male often misses and very few eggs get fertilized. This, coupled with dwindling fish stocks, is why man has intervened in this process – although fish farming is often met with a certain amount of controversy due to breeding techniques and what the fish are fed.

We stirred the milt and the eggs with a feather to mix them up, then placed the eggs in a jar and filled it with stream water to transport them. During the water-hardening phase, where the liquid fills the shell of the egg and separates it from the yolk membrane, the eggs are quite robust. But, after a period of no more than 48 hours the eggs become very susceptible to vibration; the slightest tap on the side of the box could kill them all. This meant that the clock was running and we had to get our fertilized eggs to our hatchery as quickly as possible.

fragile. This stage is dependent on water temperature. The embryonic fish then continues to grow within the egg for around two weeks before hatching out as an alevin. An alevin is a small fish that is still developing, with the orange yolk sac of the egg attached to its stomach. It uses this as a source of food for several weeks while its eyes and mouth develop. Once this stage is complete, the fish is known as a fry and will be able to swim and feed itself. The fish will reach full maturity at around two years.

Miracles do happen! Despite the cold winter, we have fish (well, alevins at this stage).

At one point the winter was so severe I honestly didn't think that we were going to get a single egg to hatch – not because I was worried that the temperature wouldn't rise but because I thought that the cold must have killed them all off. Imagine my surprise and sheer elation when one day I opened the hatchery to see lots of little electric-coloured alevins moving around in the chambers of the egg box. Success. Straight away I followed Horace's advice on feeding the fish. McNicol would prepare a meal of ground-up offal mixed with crushed dog biscuits that had been soaked in water. He also used to hang carcasses of dead animals such as rabbits in cages with biscuit tins below them to catch the maggots produced by blowflies. The smell is bad but the results are good. This varied diet would have been fed to the fish to bring them on to a point when they were ready to be released into the river, typically when they are around 4 in (10 cm) in length. I released my fish into the mill pond with a definite sense of achievement.

THE TANNERY

There has been a tannery in Hamlyns Colyton, Devon since Roman times and the one currently occupying the site is thought to be the only remaining traditional oak-bark tannery in England. Early on in the project, one of our bullocks was dispatched and we wanted to use every part of the animal including the skin. I took the hide along to the tannery to see how the process works, hoping to walk away with a nice piece of leather. During my visit I was shown the ropes by Andrew Parr.

The first stage is to remove any flesh with a knife, then the skin is soaked in baths of lime of varying strengths. This opens the pores and loosens the hair so that when the skin is scraped again using a three-bladed knife all the dirt, dead skin, and hairs are removed. The hide is then washed and cut up: different parts of the skin (the butt is considered the best) tan at different speeds due to the thickness – and the Red Ruby skin is certainly thick. These preparations strip the skin down to the central layer, known as the corium or derma.

Naively I thought that the process might take only a matter of weeks, but to tan a hide fully it takes around 15 months.

At Hamlyns Colyton they immerse the skins in various tanning baths of different strengths, containing traditional tanning liquor made from oak bark and water. In the initial tanning baths the skins are hung vertically from frames, which are rocked to and fro. To power this set up the tannery relies on a waterwheel, as it has done for centuries. The waterwheel also powers a bark shredder that turns large strips of oak bark into little chips of bark. These chips can then be added to the water to adjust the strength of the liquor. Having a go at pulverizing oak bark is one of the most satisfying tasks I have ever done and Andrew said that he never tires of it.

In the later tanning baths the hides are laid flat with ropes that are attached to a beam at the surface of the tank. The hides need to be turned every so often and when each hide is laid back into the bath, a handful of oak chips is thrown in. The hides are surprisingly heavy – but then the skin is the largest organ in the body. It is not unknown for the occasional man to fall into the baths, but as no chemicals are being used the worst that usually happens is a slight non-permanent staining of the skin.

Once the hides have gone through the baths, they are hung up to dry. During this phase they are periodically removed and oiled, using a flat-bladed knife in a pushing motion. This stretches the hides and ultimately makes them thinner. Then they are treated with a mixture of tallow and cod oil – a process known as stuffing. Some of the hides may then be dyed to order. Unlike chromium-tanned leather, which is treated using chromium salts, traditional oak-bark-tanned leather will revert back to its normal state if left untreated.

The tannery at Hamlyns Colyton used to make leather from a variety of animals, including sheep and goat (which is used for finer leathers such as chamois), but it now specializes in cow leather. I suppose my Red Ruby skin is still in those tanning baths but I took away a finished hide and Mathew Payne, a local saddler, came to make some coupling straps for the horses from it. The coupling straps join two horses together. One strap goes from one bit to the other bit. One strap goes from one bit to the opposing horses collar. The third strap mirrors this, going from the other horse's bit to the opposing collar (this forms a kind of triangle). The straps he made were beautiful and if looked after properly will last a lifetime. He recommended neat's foot oil (see page 60) to keep them in good order.

ABOVE A chart of tannery tools.
BELOW Scraping the hair and dead skin cells off the hide after it has come out of the lime baths.

Mining

It is said that the definition of a mine is 'a hole with a Cornish man at the bottom of it'. Here in the Tamar Valley where Cornwall meets Devon, the landscape has been shaped, scarred and forever altered by its mining heritage. Chimneys and crumbling engine houses mark the pit heads; long straight tracks were once incline railways; deer fencing surrounds gaping shafts. The remnants of powder houses are nestled in the undergrowth next to cracks in the rock that lead into amazing manmade underground cathedrals. Inland ports along the river such as Calstock and Morwellham Quay shipped ore from several mines to the coal fields for smelting. It was easier and cheaper to ship the ore to the coal rather than import the vast amount of coal needed to smelt the ore. On the return journey the ships carried limestone to burn in the kilns, and slate for use as a building material.

Towns grew and the population of this part of England swelled. The 7th Duke of Bedford remodelled medieval Tavistock, which was a stannary town established under King Edward 1's 1305 charter; using the mining profits, he built model cottages in and around the town such as those that can be seen at Morwellham. But the boom was short lived. As the world got smaller it became cheaper to mine copper in other parts of the globe, such as South America, South Africa and Australia. And when the London wholesale bank Overend Gurney and Company collapsed, the share price of copper dropped and Cornish copper mining was no longer profitable.

Many of the miners took their trade and culture overseas, exporting a wealth of expertise around the world. It is estimated that more than a quarter of a million Cornish workers (or Cousin Jacks) emigrated in the last half of the 19th century. Their legacy lives on in a variety of ways: processes honed in the Tamar Valley are still in use in mines today

and the world's largest Cornish festival – Kernewek Lowender (Cornish Happiness in the Cornish language) – takes place on the Copper Coast of South Australia.

Peter, Alex, Rick and Phil emerge from the mine, ochre and all.

The Duke of Bedford, who now had designs on tourism, imposed a post-mining reversion clause on the shareholders of Devon Great Consols Mine, which meant that buildings and infrastructure were cleared and the industrial landscape was planted with conifers. Ironically you can see Devon Great Consols for miles around as the land is devoid of vegetation due to the arsenic mining that occurred there; in the last half of the 19th century more than 70,000 tons of arsenic were mined. It is a popular belief that at any one time while the mine was active there was enough arsenic stored on the quay at Morwellham to kill the entire population of the world.

MINING IN THE AREA

Cornwall and Devon have a long history of mining, beginning in the Bronze Age and effectively ending with the closure of South Crofty tin mine in Cornwall in 1998. In this area many minerals have been mined, including lead, silver, manganese, tin and arsenic, but the mining boom in the Tamar Valley in the mid-19th century centred on the rich copper deposits. Between 1844 and 1902 more than 700,000 tons of copper were mined; the peak year was 1855–6 when production hit 209,305 tons. The subsequent increase in the share price of the mines in a single year earned Morwellham Quay the title of the 'richest copper port in Queen Victoria's Empire'. (The ore was shipped to South Wales for processing, giving Swansea the title of Copperopolis.)

FOSSICKING

By the Edwardian period mining in the Tamar Valley had dwindled especially on the Devonian side of the river, but it was still possible to scratch out a living by picking over the spoil heaps of previous mining enterprises. This is a process known as fossicking – the term is also used to denote prospecting in Australia and Cornwall. Lodes – rock containing ore – brought out of the mines had been dumped on the dressing floors (see below) and when the operation came to a halt they had just been left. Rick Stewart and Phil Hurley, two historical mining experts, took us to the dressing floor of the George and Charlotte mine at Morwellham Quay where the remains of the last lode to have been brought up to the surface in the mid-Victorian period lay on the ground.

The dressing floor is a flat area of land where the rocks containing the ore were smashed up using hammers before assaying to determine their mineral content and therefore their value. Generally these rocks were brought up to the surface via winches and shafts, whereas the spoil from the mine with little to no mineral content was taken in tip wagons out of the entrance of the mine and dumped or was used to create a working floor as the miners chased a lode. On the George and Charlotte dressing floor all the pounding was done by hand (usually by women), starting with a big hammer and going on to smaller hammers as the lumps decreased in size. The dressing floor was connected to the quays by a railway so the ore could be easily and quickly transported to the assayer's labs and then shipped out for smelting.

Although it was no longer profitable to mine copper, it was still a valuable metal and fossicking requires only time and effort by an individual. Copper (atomic number 29) is easily recyclable and is the third most recycled metal (after iron and aluminium): it is estimated that 80 per cent of all the copper ever mined is still in use today. Man's relationship with the metal goes back at least 10,000 years. This is because copper is one of the few metals to occur in nature in an uncombined 'native' form, meaning it is found in its metallic state with no smelting required to extract it. Copper is a ductile metal, which means that it can be stretched without fracturing – for example, being pulled out like a wire; it is also malleable, which means it can be hammered into a plate. (These are very similar terms but some metals, such as lead, are only malleable and not ductile.)

Copper's mechanical, electrical, thermal and germicidal properties mean it is used in a wide range of products. With the growth of the electrical industry in the Edwardian period there was a market for the discarded fragments of copper ore that littered the spoil heaps of the mines of the Tamar Valley.

Fossicking does require a lot of time and effort and if you have a farm to run it won't necessarily pay the bills. So Rick and Phil told us of another process used in Edwardian times to collect copper with relatively little effort.

THE PRECIPITATION TANK

A precipitation tank uses wooden 'launders' or troughs to channel water that is flowing through the mine into a wooden tank containing scrap iron – for our experiment we used iron railway sleepers from the mine. The mine water contains copper sulphate and when it comes into contact with the iron a single displacement reaction takes place, in broad terms this means that an element or ion moves out of one compound and into another – or, simply, one element replaces another.

In our tank we started off with iron and a solution of copper sulphate, and ended up with a layer of copper deposited on the scrap iron and a solution of iron sulphate. The iron and copper atoms literally swap places, very slowly. The reaction takes place because the iron atoms are more reactive than the copper ones.

Over time the iron sleepers in the wooden tank gained a thin film of copper that could be scratched off with a knife and collected up. We had to use wooden launders and tanks (which would have been sealed with pitch at the joints); if we had used iron drainpipes all the copper would have been deposited on the inside. The process is dependent on a steady yet gentle flow of water coming out of the mine, a large surface area of iron, and a good deal of patience.

We left the tank for a couple of months over winter, checking it a few times and seeing the formation of a yellowy sludge on the iron sleepers. This was later replaced by the unmistakable metallic sheen of copper. Once the tank was set up the effort involved in extracting the copper was minimal.

Edwardian miners would have been well aware of the potential to use the reaction to extract copper; it was said that the iron rungs of the ladders down the mine took on the bluish hue of copper when water dripped on them. One of the by-products of the process is ochre, which when dried out and separated, makes a good pigment for paints.

ACROSS THE RIVER

On the Cornish side of the river Tamar mining saw somewhat of a revival in the Edwardian period. There were large deposits of two different ores along with the

ALEX'S DIARY

Armed with some freshly sharpened picks, an extremely large hammer, a drill bit and a basket full of pasties, we delved into the heart of the Devon hillside. It wasn't long before I lost all sense of orientation as we made our way along tunnels and up and down various shafts. We were mesmerized by vast caverns opening up above our heads where, over a century and a half ago, the rock of the hillside had been painstakingly hewn away by hand. When it came to our turn to drill out some rock I was surprised at just how hot it became in the cramped conditions and very quickly started to peel off the layers. As we worked away the air became stuffier and a feeling of claustrophobia was soon to descend upon me.

Exhausted by the hand drilling, we then took up the pneumatic drill – a feature of Edwardian mining – and with the dust kicked up from this machine I felt my chest getting heavy. It dawned on me very quickly that I was a man of the fields and I soon began craving the open spaces. I had enjoyed our mining experiences enormously but I can't tell you how relieved I was to find myself back out in the fields the next day setting the plough for another day spent behind the horses.

copper. Wolframite is the ore from which tungsten is extracted; cassiterite is the ore that yields tin. They often occur together in nature. Wolframite has a specific gravity of 6.89–7.1 and cassiterite 7.0–7.5. This overlap meant that it was not possible to reliably separate the two ores until John Price Wetherill (1844–1906) invented his Magnetic Separator, which proved extremely useful in Cornwall.

After heating, the raw ore was fed into the separator, on to a moving belt that passed by two electromagnets. The first electromagnet was weak and removed any iron deposits from the belt; the second electromagnet was strong and removed any wolframite, which is only weakly magnetic. Suddenly it was possible to process all the ore that had been tipped on the spoil heaps. Lots of fossicking went on and there was renewed emphasis on mining the previously abandoned lodes. Tin was a valuable metal with increasing applications in soldering and in the Edwardian period tungsten (which has the highest melting point of all non-alloyed metals) began to be used as a filament for electric light bulbs and to strengthen cannon shells.

GOING UNDERGROUND

Rick, Phil, Alex and I formed a mining 'pare' or team and prepared to go underground to chase a lode. Looking at the Tamar Valley it is hard to comprehend the complexity of the mine systems that honeycomb the hillsides. Mines were formed on many levels and the shape of the mine depended on the ore that was being extracted. Unlike collieries, Cornish mines were renowned for not having straight shafts like collieries. Therefore, Cornish expertise in mining was in demand, as they were adept at constructing belt and pulley systems and chutes (often called mills or cousin jacks) that operated in this irregular environment.

On my first experience I went with Rick to check out a mine that hadn't been entered (other than by a few caving enthusiasts) since its closure in 1869. We had to wade through waist-high ochre (which temporarily stains the skin – in my case, orange) and, in places, neck-high water until we reached the main chamber. The sight that greeted my eyes was a cavernous opening entirely hollowed out by hand drills. It must have been some 50 feet (15m) high and about 10 feet (3m) wide. At the base were all the wooden props supporting the hanging face and high above our heads were the wooden platforms where the miners worked. We pressed on underground for a distance of three quarters of a mile (1.2km) along a tunnel that had been established to find new lodes. Occasionally we would pass a short off-shoot and the ground beneath

the water was littered with abandoned sleepers. The air got thinner as we progressed because the rear shaft had collapsed. We were greeted with amazing stalactites and stalagmites that had formed from the ochre over the past one and a half centuries, and the shining luminous beads of bacteria that adorned the tunnel walls.

The mine where we had a go at drilling was a lot easier to access and a lot larger, with multiple levels and a waterwheel for pumping out the liquid from areas below river level. The mines stretched for long distances underground in all directions. It is one thing navigating the small passages between levels on a single visit, but having to do it every day to get to work carrying heavy and cumbersome tools in the near dark must have been hard. Or so I thought until I started drilling.

Miners were issued with candles but in order to save resources they would have only one candle lit for every two to three men as they navigated the tight tunnels. Once we got to our drilling area we fixed our candles to our hats and to the walls using clay. When underground in the pitch black it is amazing the amount of light candles do give off.

DRILLING

Once they had found a lode – a body of ore in the surrounding rock – miners would work it and follow it until the ore ran out. A series of shot holes would be drilled and then packed with explosives, to blast a face before starting the process again. Shot blasting was pioneered in Hungary in the 17th century and was soon in use in Cornwall. Straws or quills filled with black powder were used to ignite the charge until safety fuses were invented in the 1830s by William Bickford of Tuckingmill. The safety fuse meant fewer accidents.

During the height of mining in the Tamar Valley the shot holes would have been drilled by hand. A hand drill is essentially a metal bar with a hardened chiselled end. One miner held it while another struck it with a hammer. Between blows the miner turned the metal bar. Miners prided themselves on being able to hit the drill at any angle in any direction and using either hand.

When Alex and I had a go I was holding the drill. The thing I found the hardest was turning the drill bit: I had to pull it out slightly before twisting it round, while keeping the end of the drill in the same position for Alex to strike. When the miners drilled at a downwards angle the dust produced would mix with the moisture in the air and create a mortar. The space we were working in, although chilly and dry when we arrived, rapidly heated up with the warmth given off by our bodies, lamps and candles – and the moisture from our breath condensed on the ceilings and the walls. This was only the merest glimpse into the constant hardship that this life entailed and our drilling was pitiful compared to the inroads the miners made. They could do several inches in a matter of minutes.

Using a pneumatic drill in such a small space is quite an experience. The noise, the dust, the vibrations and the sheer power are something that is hard to imagine.

By the Edwardian period compressed-air reciprocating pneumatic drills had come into use in mining. R Stephens & Sons Climax Imperial bar and arm drifter was in use from 1896–1914 and was, according to Rick 'possibly the best rock drill ever made'. This is because there was no attempt to deal with the dust produced – other models used dust suppressors – which earned it the nickname the 'widow maker' because it could seriously reduce the lifespan of miners who used it.

ASSESSING THE ORE

To see how our ore would have been processed in the Edwardian period we went to the King Edward Mine (as it was named in 1901) that was acquired and operated by the Camborne School of Mines from 1897. Here they had all the latest machinery of the age in use in the Cornwall mining industry with Californian Stamps (used in the gold mines in America but based on the Cornish stamps that were used in tin mining) that pound and crush the ore and a variety of machines such as the shaking table that separates out the desired minerals from the other particles.

However, before a mining pare's ore could be processed its mineral content and, therefore, its price, had to be ascertained. A sample of the ore was crushed by hand mimicking what the stamps would do to the main body of the produce from the mine. This powdered sample was then placed on the vanning shovel along with some water.

The vanning shovel was a long handled tool with a flexible spaded head (similar to a Devon shovel) upon which the powdered sample was expertly flicked. This would draw out the desired minerals and indicate the percentage that the ore contained. The main body of ore was then processed on implements such as the shaking table and the Frue Vanner (of which the King Edward mine claims to have the only working example in the world). The assessed mineral content of the ore was always fractionally less on the vanning shovel than it was in reality thereby ensuring a favourable profit margin for the mine. Once the mineral was separated out form the other particles it could be smelted and cast into ingots but by this time the miners were hard at work back underground.

Accuracy and honesty to the highest degree were required of the assayer.

SUPERSTITION

The miners would break to eat their pasties at 'croust' time. Invariably they would sit on benches rather than the floor – a small comfort in a hard world. The pasty would be wrapped in a cloth. If kept close to the body it could stay warm for several hours and represented a very balanced meal. There is a superstition that the crust of the pasty was thrown to the knockers (beings that inhabited the deepest and most remote parts of the mine and knew where the most profitable lodes where), but Rick didn't put much faith in that. He did, however, ban us from whistling down the mine, saying that it would draw the attention of evil spirits and chase away the richest ores.

Cornish Pasties

After a service at our local Methodist chapel, I got talking to several women in the congregation about food. It was soon clear that I was going to need a master class in pasty making. Everyone seemed to feel so strongly about the subject – a matter of local pride.

The boys were due to go off on a mining expedition and I would need to provide food for them so it seemed a good opportunity to get together in the cottage kitchen and produce a giant batch. We had a really fun afternoon, gossiping away as we cooked and I learnt such a lot about the local Tamar Valley food culture.

METHOD

INGREDIENTS

For the pastry:

1 lb (450 g) plain white flour

8 oz (225 g) lard

Water to mix

For the filling:

8 oz (225 g) beef skirt

1 medium onion

2 large potatoes

Half a swede or turnip

Salt and pepper

Begin by making the pastry. Sift the flour into a bowl and rub in the lard until the mix resembles breadcrumbs. Slowly add the water – the colder the better – stirring it in with a knife until the mixture begins to stick together in large lumps. Now collect the pastry all together with your hand and gently work into a ball. Cover the bowl with a damp cloth and put to stand somewhere cool. A fridge works well; I stood it in the draught by the sink.

Now chop up your beef. Skirt is a cheap cut that is a bit too tough for frying but works well in pasties if cut up small.

Peel all the vegetables; they now need to be 'chipped'. Not just chopped. All the local women were quite adamant about this. Chopping left the vegetables in too thick a chunk to cook through properly within the pasty, but 'chipping' turned them into thin slithers that would quickly become cooked through and soft. The vegetables were chipped in the hand rather than on a board. I think you could achieve something similar using a potato peeler.

Toss the meat and veg together in a large bowl, clear the table and retrieve the pastry from its cold spot.

Roll the pastry out and cut circles – around a dinner plate gives a good size. Place a generous portion of the filling on the pastry somewhat to one side and fold the pastry over the top. Sealing the pasty is done by folding a small triangle over at one end with one hand and pressing it firmly down with a finger from the other, and then begin folding the next bit of edge over in a triangle and so on till the pasty is sealed.

For Cornish pasties it was generally agreed by those who had come to give me my master class the seal or crust should be flat against the table, while for Devon pasties the crust ran over the top of the pasty. No one could think of any particular difference in the recipes themselves, it was all in the position of the crust.

Put the pasties on a lightly greased baking tray and bake in a pre-heated oven, 425ºF (220ºC/gas mark 7) to begin with. Turn it down to 325ºF (170ºC/gas mark 3) after the first 15 minutes. Bake for a further three quarters of an hour or until golden brown.

Alex and Rick munch their pasties; Croust time was normally taken sitting on a bench – a little comfort in a hard world. The pasties must have gone down a treat – not a crumb came back out of the huge batch we made.

LIME KILN

All along the banks of the river Tamar are sets of lime kilns, which were last in use in Edwardian times. Lime is one of the most widely used and versatile historic building materials and we are seeing a resurgence in its use today. It is mixed in mortars, plaster, lime wash and beaten earth floors; lime breaks down organic material, preserves food and is spread on the fields to prepare the earth for some crops.

During our time on the *Victorian Farm* we had used our fair share of lime in building projects but we had never touched on how lime was made, so when the opportunity arose we jumped at the chance. Our first step was to find a set of lime kilns where we could have a lime 'burn' – the process of changing the chemical composition of limestone. We had hoped to use the kilns at Morwellham, but a mixture of red tape and their sheer size stacked the odds against us, so we went to a site on the Wenlock Edge in Shropshire.

These kilns had four chambers in a row and like many other kilns they were built into the hillside. This enabled the lime and the coal to be loaded into the top using a railway that ran along the hill. Having four chambers meant that the kilns could be constantly producing lime. At the same time the latent heat would be retained in the fabric of the building, so, when the burn was started, the energy released from the coal didn't need to be spent on drying out and raising the temperature of the kilns.

We did not have a railway and our kilns were cold, wet and damp – we could tell this was going to be an uphill struggle in more ways than one. But we did have Colin Richards MBE SCO NT ARGG[1] and with him on board we knew we could do anything. We had worked with Colin when filming the *Victorian Farm* Christmas 'specials' and knew he was the man for the job.

OPPOSITE The kilns at Morwellham Quay.

[1] *Member of the British Empire (awarded for services to conservation), Shropshire Conservation Officer, National Treasure, All Round Good Guy.*

THE LIME CYCLE

The lime cycle is a simple and somewhat miraculous process. It starts with rocks of calcium carbonate ($CaCO_3$), such as chalk or limestone (also known as lime). You heat it to approximately 1,652°F (900°C) – if you don't heat it enough calcination will not take place, heat it too much and you will be left with un-reactive dead burnt lime.

The result of the heating process is that all the Carbon Dioxide CO_2 has left the Calcium Carbonate $CaCO_3$ leaving Calcium Oxide CaO (also known as quick lime, or lime). This will weigh approximately half what the initial chalk or limestone weighed. This Calcium Oxide CaO is very reactive and is added to water H_2O in a process known as slaking. The result is Calcium Hydroxide $Ca(OH)_2$ (also known as slaked lime, hydrated lime, or lime). The Calcium Hydroxide $Ca(OH)_2$ is then used as a component of things such as mortar or plaster which is used in buildings. There is then an exchange process of carbonation where the water H_2O evaporates from the Calcium Hydroxide $Ca(OH_2$ and Carbon Dioxide CO_2 is absorbed. The result is Calcium Carbonate $CaCO_3$ which is the same material as we started with such as chalk or limestone (also known as lime).

The amount of Carbon Dioxide CO_2 absorbed is the same as the amount lost during the heating process. This makes lime a very attractive building material from an ecological point of view coupled with the fact that it takes less energy than cement to make and it allows the buildings that it comes into contact with to breathe.

THE BURN

This is all very good in theory but we were still left with the daunting task of undertaking the burn. The chamber we were using was somewhere between conical and cylindrical in shape, open at the top and with a small oven-sized doorway at the bottom for fire lighting and unloading. There was also a hole for riddling (poking the load with a stick should it become stuck) and observing the burn. On larger kilns, such as those at Morwellham, there are more riddle holes further up the kiln. These often sit alongside with recesses referred to as bread ovens (lime burners used them to bake bread and other food).

To load the kiln we built up a base of wood as a kindling to ignite the first layer of fuel. We added a layer of coal followed by a layer of limestone (in this case we were using chalk), followed by another layer of coal and another layer of limestone and so on and so forth.

For the burn to be successful the kiln needed to be at least two thirds full. We loaded it with 12 tonnes of limestone and 3 tonnes of coal all added by shovel. Stafford, the fourth member of our team, asked us to keep count of every shovelful. He was making

When Peter expressed a burning desire to undertake a lime-burn, I almost fainted with disbelief! After the long weeks spent plastering, whitewashing and mixing mortar for stone walls back on the Victorian Farm Peter's exact words had been, 'If I never see another bucket of lime, I'll die a happy man.' Lime, even in its hydrated form, is a really rather caustic substance and after lengthy exposure to it, our hands and faces had smarted and cracked from dryness. And yet, here he was, at the beginning of our Edwardian Farm adventure waxing lyrical about a process that was infinitely more dangerous to our soft 21st century complexions than simply working with the stuff. A 'lime-burn' would see us closer than ever before to a substance that, before re-hydration, reacts so violently to water that even the slightest contact with a sweaty forehead or exposed arms can cause severe burning. A sample the size of a pea, if it came into contact with the eye would burn out the whole socket in a matter of seconds.

So it was with trepidation that I had agreed to help Peter with this endeavour and now, some six months on, I was beginning to see the benefits. Our primary objective with the burn had been to produce enough quick lime to fertilize our fields for our desired arable projects, but we had set some aside for any other projects that might require lime. As the year has progressed I was starting to see what a wonder material it is. Not only has it done a wonderful job of setting our oat crop on its way, we have used it for sterilizing the milking parlour, fending off pests in the market garden, dissuading a host of nasty insects in the poultry houses and as a base for a mortar for minor building projects. I have even experimented with preserving eggs in a mixture of water a lime powder. In short, it is amazing stuff and it has been useful to so many areas of our farming year that I can't thank Peter enough for pushing ahead with its production so early in the year!

a detailed study of the burn and its results. He also asked us to keep the coal away from the sides of the kiln as, although it was lined with firebricks, it was also quite damp and they might explode if they got too hot too quickly.

The ratio of coal to limestone varies, but it needs to be around 1 part coal to 5 parts limestone.

Shovelling 15 tonnes of material by hand into a hole took a fair effort. Luckily, Alex, Colin, Stafford and I had a fifth team member, Mick, and when the cameras weren't rolling, our crew – Stuart Elliott (series producer/director) and Mathew Nicholson – pitched in. We worked well into the night and the job satisfaction of having the thing loaded was immense. Once we lit the kiln there was no going back; it either burned or it didn't. We sincerely hoped that it wouldn't go out with the first match, but equally none of us expected every layer of coal to burn fully and all the lime to be turned into highly reactive calcium oxide. We would just have to wait and see.

Mick had erected a traditional typical temporary shelter on the outside of the kiln, consisting of deadwood from the forest and odd pieces of corrugated tin. It gave us somewhere to shelter if it rained and a dry area to unload the volatile quick lime at the end.

When it was time to light the touch paper we all huddled in Mick's shelter at the base of the kiln. I don't want to use the word anticlimax but it was not remarkable. The fire is contained and so it is not really visible to the outside. For a long time we were just hoping that it was doing what we wanted it to do. After a while smoke began to pour from the top in immense thick white-grey-yellow-black clouds, reeking of sulphur. It was a mixture of emissions from the coal and the 6 tonnes of carbon dioxide that would be produced. There were four kilns and we had loaded just one. What must it have been like to load the other kilns while burns were going on.

Obviously the workers would have been tipping the stone and the coal in, rather than shovelling it in but it would have been horrific conditions. There are accounts of tramps sleeping at the kiln openings for warmth. Sometimes these gentlemen of the roads would be overcome by the fumes and fall into the kiln. They would not come out.

UNLOADING

The burn took three days and once the kiln was underway there was little we could do. To keep the rain off, temporary shelters were often erected above the kilns and turfs or a layer of clay placed on the load to dampen the flames and control the rate of calcination. We used sheets of tin to smother the skyward flames coming out of our kiln. Meanwhile, Alex cooked us a fantastic spit-roasted rabbit with a vegetable and bean stew and Colin as ever had brought a good supply of beer. We ate and drank in another canvas and A-frame shelter typical of the period and thought of those who had once done this as a living.

Our burn was a total success with all of the limestone being turned into quick lime. All that was left was to shovel it out and sort it. We did this by putting it through a grill so that the larger, better lumps went into barrels that were to be slaked to make lime putty for building projects, and the smaller particles that contained ash and cinders went into barrels destined for the fields (the ash and cinders would help fertilize the earth). The lime was hot and we realized that the burn was still going on, so this slowed us down slightly. While I was shovelling out the quick lime I was grateful that we hadn't gone for a larger set of kilns. The ones at Morwellham are about twice the size – the largest on Wenlock Edge are 100 feet (30m) deep.

THE RESULTS

Everyone was impressed with the results. Colin told us that the lime had a shelf life of a week or so – after this the moisture in the air would make the quick lime un-reactive. We spread most of our lime on our fields to slake in the rain and to help neutralize the acidic soils of this area. The rest of it we slaked ourselves. The easiest way to do this was to dig a pit, fill it with water, then add the lime. I had heard about the violence of the reaction but it was only when I saw the water boil (and later the canvas we covered it with catch alight), that I appreciated the true nature of this substance. The initial reaction is intense but it calms down after a while and will continue to slake ad infinitum. The longer you leave the lime before using it, the better it is: lime putty in barrels was often given as a wedding present and handed down through generations because lime will stay good in a barrel for a thousand years or more. However, if you use lime too early (after a month or so) in, say, a mortar, it will continue to slake in the wall which will result in blow out – little pock marks in the pointing. This is fine for makeshift agricultural buildings but not so good for structures that you really want to stand the test of time. After the initial slaking I put the lime in a barrel for use by future generations.

THE DANGERS

Unloading was the process that I was most apprehensive about, not because I was worried whether it had worked but because the 7 tonnes or so of calcium oxide or quick lime that was going to have to be shovelled out of the bottom has to be one of the most naturally caustic substances known to man. It will react exothermically with any water it comes into contact with, so if it touches the sweat on your skin generated by shovelling it or, heaven forbid, you get it in your eyes, it will burn or blind you. I had made myself a head cover and gloves but needless to say I got it on my skin and in my eyes – luckily only in dust form so I could easily wash it out, but it gave me another insight into the life of a lime burner.

FINCH FOUNDRY

A WATER-POWERED FORGE

The role played by water power in the industries of the south west is nowhere better illustrated than at Finch Foundry in the village of Sticklepath, on the northern foothills of Dartmoor. Water power has been harnessed on this site since medieval times, when the economy of Devon was founded on the wool industry. Water wheels powered both grist and fulling wheels. A grist mill very simply ground down cereals crops into flour or animal feed, while a fulling mill beat out the oils and dirt in wool while at the same time bulking it out, using a series of hammers called fulling stocks.

The story of the foundry begins in 1814, after the collapse of the wool industry, when a family of blacksmiths from Tavistock leased the woollen mill and converted the waterwheels to power a forge. The Finch family went on, in the 1850s, to lease the old grist mill and used the waterwheel there to drive an enormous grind stone. They also set up a saw mill and a cartwright's workshop. In short, if you needed something for your farm made from either iron or wood, Finch Foundry was the place to go.

The grind stone sharpened tools quickly and effectively, so that the Finches could produce an enormous range of 'edge' tools with a sharp working edge – it is these they are most famous for. During the course of our work we used a number of examples, from hedging hooks, pleachers, grass hooks and scythes to cleavers, spar hooks, adses, axes and rick knives. When I was hedging I envied the bill hooks the rest of the team were using. The bill-hook is the hedger's primary tool and because hedges all over the country used varying species of shrub and were 'pleached' or laid in different ways, regional variations in bill-hooks arose. I needed a Devon-style bill-hook.

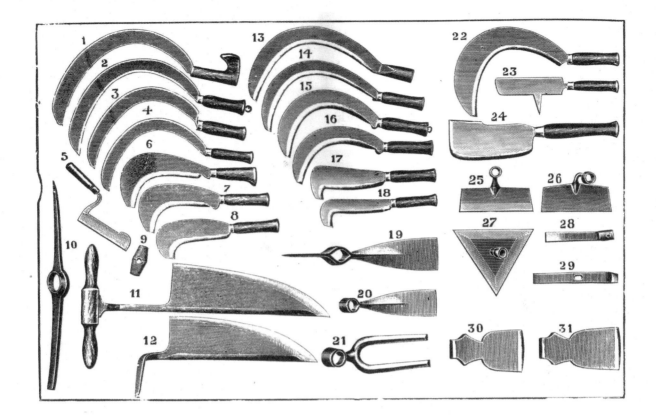

BLACKSMITHING

I enlisted the blacksmithing skills of Simon Summers – a man of extraordinary talents – and we headed over to Finch Foundry. Roger Boney, custodian of the foundry showed us the water-powered technology that had meant this foundry could produce over 400 tools a day when it was a going concern. First we set in motion a small wheel to power a fan box. A constant supply of air jetted into the hearth is the only way to achieve the high temperatures for melting metals – in traditional foundries this was usually supplied by a huge set of hand-cranked bellows. At Finch Foundry, however, a paddle-like fan in an enclosed box pushed air into a pipe that ran through the floor of the foundry and supplied five hearths with a jet of air. This was the first major saving in manpower. The second was the giant trip hammers powered by the largest waterwheel.

Of the two hammers, the biggest was the plating hammer, weighing in at over 20cwt (1,000kg), and used to beat iron into plate metal for shovel blades. The steeling hammer was slightly smaller at 15cwt (726kg) and this was used to forge weld wrought-iron with blister steel to make blanks for edge-tool production. The steeling hammer strikes at 240 beats per minute and in doing so generates heat in its own right. Consistent heat is needed to forge weld two pieces of metal together. Finally, an

The range of 'edge' tools produced at Finch foundry included bill-hooks, spar hooks, grass hooks, axes, hoes and rick knives amongst other must-have farm tools.

enormous set of shears were powered from the same wheel, to cut red-hot bars of metal up to ½ inch (13mm) thick. Understandably, our use of the hammers would have to be limited. While Roger was keen to see them used once more, their unique status as England's only set of working water-powered trip hammers needed to be maintained. Anything more than a speedy demonstration ran the risk of damage – not just to the hammers but also to the buildings around. Famously, the neighbouring public house had to fix a rim to its shelves to stop the heavy vibrations from the hammers sending the glasses crashing to the floor, and in 1960 the working life of the foundry was brought to a premature end by the collapse of the rear foundry wall.

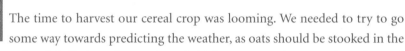 REPOUSSÉ WEATHER VANE

The time to harvest our cereal crop was looming. We needed to try to go some way towards predicting the weather, as oats should be stooked in the field for two church bells in order to ripen (which means cutting on a Saturday and bringing in eight days later on the Monday). Other than reciting folklore, watching swallows hunting for flies and nailing seaweed outside the cottage, we had little to go on. So Alex decided to make a barometer and I went to master blacksmith Simon Summers to look into making a weather vane. At least we'd know which direction the wind was coming from, an indication of when we were in for a good spell of weather.

The first recorded weather vane is the effigy of the god Triton who was mounted on top of the Tower of Winds near the Acropolis, dating to around 68BC. Weather vanes on top of churches often take the form of a cock, as this relates to St Peter denying Christ three times before the rooster crowed. The oldest weather vane currently in England is thought to be at Ottery St Mary in Devon (dating from 1335) and it is equipped with tubes that whistle in the wind to imitate a crowing sound. However, weather vanes were around earlier: the Bayeux tapestry depicts a craftsman attaching a weathercock to the spire of Westminster Abbey. (The word vane is thought to come from the Anglo-Saxon word 'fane', meaning flag and weather vanes may have been used as replacements for flags that once adorned the tops of churches).

Simon wanted to make a weathercock using the techniques of repoussé and chasing, which is similar to embossing. Like many crafts it was an art form that was waning in the Edwardian period with the advent of mass-production techniques and the loss of a generation of practitioners in the war. Craftspeople like Simon are keeping the techniques alive today. Repoussé has been used throughout history to produce amazing works of art from metal. Two of the most famous pieces of repoussage are the death mask of Tutankhamun and the Statue of Liberty. In this country one of the most revered practitioners was the Frenchman Jean Tijou. Tijou is thought to have been a Huguenot who trained in Versailles and came to England at the start of the reign

At 726kg, the steeling hammer was a quarter of a tonne lighter than the plating hammer and struck at 240 beats per minute.

of William and Mary (1689). He stayed for 20 years, producing the metal work in buildings such as St Paul's Cathedral and Hampton Court Palace, before mysteriously leaving – having never been fully paid – around 1710.

Making the Weathercock

Repoussé means 'to push back' and is the technique of hitting metal from the underside. This, coupled with chasing (which is working metal on the front face), will produce a 3D design from a flat sheet of metal. The first step is to create a flat sheet of metal. This is done by taking an ingot of the metal to be worked (in our case copper) and hammering it flat. The ingot is heated in the fire and the blacksmith (in this case, Simon) hits it with a standard ball-pin hammer and his assistant (in this case, me) strikes the same spot with a heavy sledge. The blacksmith concentrates on how to hit the metal in order to best flatten it and the assistant provides the power – the ultimate remote-controlled machine! I asked Simon who he thought was best; me or Alex. Simon said that Alex had asked him the very same thing and he would answer me when I wasn't waving a heavy object so close to his head.

ALEX'S DIARY

I had to play the part of the steeling hammer. There was, of course, little chance of me achieving 240 beats per minute but, more importantly, I had to ensure that I could bring an enormous sledge hammer down on the exact spot where Simon was striking the red-hot metal. Maximum concentration was vital as I worked alongside Simon in what is termed a 'smith and strike' partnership. Failure to be accurate would lengthen the process and failure to keep in rhythm could result in either a clash of hammers or – disaster – my hammer hitting Simon's hand.

I loved every minute of it. It was great to be involved in such a fascinating process. Simon would get the metal up to temperature, pull it out and place it – all at speed – on the anvil and strike where I had to hit. I would then bring my hammer down on that point and so the rhythm began and continued until the metal had cooled too much to work, when it was placed back in the fire to heat up again. Steadily, through our smith and strike partnership, we drew out the 'blank' and began to curve the end and, before my eyes, my bill-hook started to take shape.

The sheet is then annealed by heating it in the fire and quenching it in a trough to make it malleable. It is placed on leather cushions filled with sand, as it is pounded with a series of different blunt chisels to tease out the design. It is a noisy process. The sound changes as the metal becomes work-hardened and then it is time to anneal the metal again. Once the copper sheet was pounded flat it was time to etch the design for the weather vane on to it. Bear in mind the final result will distort the image: things like eyes in a face will be drawn further apart than they will eventually end up.

To work the other side with the repoussé technique, the indentation is filled with pitch and when this begins to cool and solidify, the piece is flipped and placed into a tray filled with melted pitch. The liquid spills over the side of the copper sheet and holds the whole piece in place. The latent heat from the pitch helps keep the copper soft and makes it easy to work. When the copper becomes work-hardened, the pitch gives up the metal, releasing it from its grip. Then it is time to reheat it in the fire. This process goes on until the relief has been perfected. For a weather vane there is less detail but it is more pronounced as it needs to be seen from a long way off. For a piece at ground level there will be more subtle detail. One thing to be wary of is that as the metal is worked it becomes thinner and easier to punch a hole through it. When the main body of the cock was completed we soldered it together using tin. The free feathers were riveted on, which would give the rooster character. It is then customary to cover the piece in gold leaf – a gentle process: one wrong breath could result in a shower of fine fleeting precious particles disappearing on the wind.

For the readers of *Swallows and Amazons* and *Swallowdale* the characters of the charcoal burners within the British landscape will be familiar. I wanted to do a charcoal burn to make a fuel that could be used to work the iron at a higher temperature, with fewer impurities than coke. Charcoal burning is a dying skill. But, by the turn of the century, the world was taking on a different aspect. Now we have mass marketed items such as mobile phones, but not a single person on this earth knows how to make one from scratch, as each different component takes so much knowledge to produce.

For our charcoal burn we called upon the services of Colin Richards who helped us make lime. He showed us how to construct a charcoal clamp and smother the flames so that the wood smouldered rather than burned. We watched the smoke, waiting for it to

Peter, the apprentice, and Simon Summers, the Master Blacksmith and Millwright.

turn from white to blue, which would indicate that we were then burning charcoal and the process was finished.

He also showed us how to smelt iron in a small furnace. This is an ancient technique that was still being employed in areas where a foundry was not easily accessible. We used this iron to create the frame of the weather vane with the cross section indicating north, east, south and west. We painted the iron with ochre, which is found in abundance in these hills and was used to dye sails and create paints. It is also one of the first materials discovered to prevent wrought iron from rusting. However, to attach the copper cock to the iron framework we needed to use a piece of wood. Otherwise a current would have been created between the iron and the copper, and the joint would have begun to corrode.

Once we finished the weathercock we erected it in our farmyard so that we could tell from which direction the wind was blowing. It also reminded us of the man hours that went into creating these enduring pieces of art that adorn many high points in our land and the craftsmen who made them.

LACE MAKING

Honiton in Devon had been an important lace-making centre for centuries and although the industry was in decline in Edwardian times, a number of women still supported themselves and their families by hand-making bobbin lace. Pat had promised to teach me some lace making.

The main problem for the lace makers of Devon was the increasing popularity of machine-made lace. Machine lace was many, many times cheaper and increasingly good quality. More and more people were able to afford it and this had reduced the exclusivity of all lace. At one time it had been a serious indicator of wealth – an investment that held its value much like good jewellery.

With so much cheap machine lace available, you had to have an educated eye to discern the expensive handmade luxury goods from the rest. Lace as a brand was suffering from cheap knock offs. Some people did still appreciate the difference: the royal

These postcards show the Honiton lace dealer shops as they were just after the great war. Their Edwardian predecessors still offered a few new pieces for sale as well as repairing and cleaning older heirloom pieces.

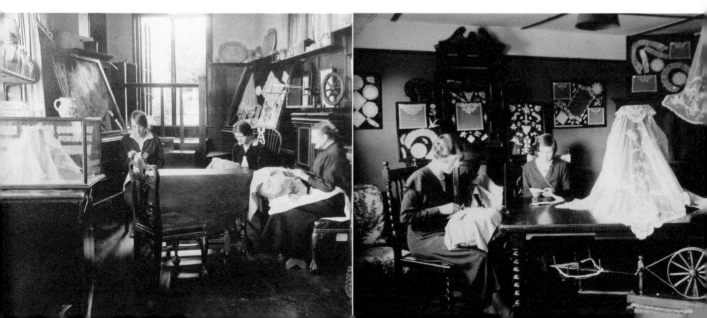

family in particular made a point of buying Honiton lace and encouraging their circle to do likewise.

While they were making one of the most expensive of all consumer goods, the lace makers themselves received a pittance. Most worked for dealers, who collected together lace from many different makers and assembled the pieces in their shops. Honiton lace begins with a series of small discrete motifs; these are then carefully sewn together to make a larger finished design. It was common for a lace maker to specialize in one or two of these small motifs – sticking to repeating the same small piece over and over again allowed a worker to become very fast and so increases both her productivity and her income. Granny Pinn, Mrs Widger's mother, specialized in a butterfly design.

This specialization made the lace makers all the more dependent upon the dealers and also left them vulnerable to changes in fashion. Learning new patterns was a risky business. A poorly executed piece could not be sold and speed would certainly suffer for some time as the worker adapted to the new design, which meant a cut in income for several weeks. Lace makers in general lived so close to the edge that they were often not in a position to take such a cut.

Most women learnt their skills when they were very young, lace schools offering tuition and discipline to children from four or five years of age. When the education act made schooling compulsory for all children, there arose a conflict of interests between the legal duty to educate children in the three R's and the local desire for children to

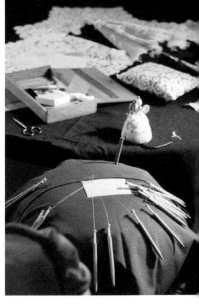

The tools of the trade are both simple and cheap – they needed to be as lace making was so very badly paid.

CATTERN CAKE

INGREDIENTS

8 oz (225 g) plain flour

1 tsp bicarbonate of soda

8 oz (225 g) sugar

1 oz (25 g) ground almonds

1 oz (25 g) currants

1 tsp ground cinnamon

8 oz (225 g) butter

1 egg

St Catherine was the patron saint of lace makers. St Catherine's Day is the 24 November and in many parts of the country lace makers made special cakes to celebrate. This is an Edwardian recipe for cattern cake, cattern being a corruption of the name Catherine. The presence of bicarbonate of soda is a real give away.

METHOD

Mix together all of the dry ingredients in a bowl. Pop the butter into a small pan and gently melt. Do not allow the butter to boil as this makes it separate. As soon as the butter is just liquid pour it into the dry ingredients and work up into a paste. As soon as it begins to come together, beat in the egg. You will have a fairly stiff paste which can be shaped quite easily.

The traditional shape for cattern cakes was wheel-like, in memory of St Catherine's symbol of martyrdom. Take small pieces of the paste and roll them out on the table into worms. Bend your worms around into circles and pinch the ends together to make them hold. Shorter worms of paste can be laid across the circle to create the spokes of the wheel.

Lay the finished wheels on a lightly greased baking tray and bake in a hot oven at 400ºF (200ºC/gas mark 6), for about 10–15 minutes or until they begin to brown.

learn how to make a living at lace making. A local compromise saw Honiton's girls receiving half a day's academic schooling and half a day at lace making.

My own attempts at lace making were woeful. Pat was very kind about my efforts, but it was soon obvious to me that I have no natural flair for it. Learning was going to be a slow process, and not one that I was sure was really worth the effort. As an Edwardian farmer's wife I needed to be able to use my time to generate a little extra income, rather than indulge in a hobby. The lace that Pat showed me, however, was so much more interesting after I had had ago at making some myself. Even a couple of hours under Pat's watchful eye and I was seeing things that I hadn't spotted before.

If I wasn't going to be able to take up lace making in any financially sensible way, maybe I could still find a way of tapping in to the local industry. Several lace shops and dealers in the early 20th century found that while they were selling much less lace than before, there was still a market in the repair and cleaning of old lace. It is skilled work and requires good eyesight, excellent needle skills and an understanding of the original lace-making techniques. The cleaning of lace is also a skilled business. Handled roughly or doused in the wrong chemicals, the lace can be easily ruined.

OPPOSITE Pat was extremely patient in her teaching.

RUTH'S DIARY

Pat sent me an old lace valance to clean. It wasn't in too bad a state: there were one or two almost microscopic tears here and there in the machine-made net backing but the handmade Honiton motifs were in exceptionally good condition. There were no major stains either, so there would be no need to employ one of the many specialist cleaning recipes. It just needed a careful wash.

This was something I felt confident in doing. My Welsh grandmother used to tell me how she did it when she was called in to help spring cleaning in some of the grander homes in Cardiff. My mother, too, remembers being taken along as a young girl.

First the lace has to be supported during the washing. I laid it on a clean linen cloth and carefully wrapped it up into a parcel. I tied the parcel with string and placed into some plain lukewarm water to soak.

Next I took the parcel out and gently squeezed to remove the excess water. I mixed up some grated soap with hot water until the soap had thoroughly dissolved and then added some more warm water. The parcel of lace went into the mix and was gently sploshed about, allowed to soak and then sploshed about some more. I carried on doing

this for a couple of hours on and off before draining and rinsing.

I squeezed out the excess water. I didn't want to risk putting the parcel through the mangle so I used my rolling pin to force out as much as possible.

I could have added a little thin blue starch to the rinse had the lace been meant to hang flat as a curtain, but as this was a valance that would be used most likely in soft folds, I left it out.

Now came the slowest part, pinning the lace out very precisely on a carpet – or, in my case, a couple of wool blankets tacked down on the bedroom floor. The edges had to be pinned out first so that the whole piece would dry square, then each motif was carefully eased out to ensure it held its intended original shape and pinned down at strategic points. I don't know quite how many pins I used, but they half filled my Coronation mug.

Once it was dry I unpinned the lace and folded it back into its packing paper.

Slow work but strangely satisfying. This lace after all had taken more than a year of someone's life to make!

THE END OF THE YEAR

Reflections on a Year spent on an Edwardian Farm

RUTH

What a year it has been! There has been so much to try and fit in. With no one thing forming the basis of our livelihood we have been rushing from one project to the next. A little tourism here, some farming there, daffodils and strawberries, mining and salmon netting, lace making and shrimping, we have had our fingers in as many pies as we have eaten pasties.

I have found so much that inspires and intrigues me in our Edwardian year, and enjoyed so many memorable experiences. Even trying to pick out the highlights is no easy matter.

The food quickly became important to me. Devon and Cornwall have such rich culinary traditions, and although we did have a number of indulgent costly meals, it was the more everyday working-class dishes that I really enjoyed both cooking and eating the most. Eeking out a limited set of ingredients and tiny food budgets is a challenge, but one greatly helped by sticking to the tried and tested meals of local tradition. The nature of our winter fare was tough at times with its endless round of bread and potatoes, but also served to highlight just how wonderful our local summer gluts were. The cycle of daily meals is perhaps one of the most evocative memories of Edwardian Devon. I know that the recipes and skills are something that I shall be using in my own life.

Ta-da! Edwardian life has been a combination of hard work and a fantastic time. I have been so lucky. Thank you to everyone who made it possible.

Like many people in modern Britain, members of my own family were in service within living memory. Some of the most emotional moments for me this year centred on domestic service because of this; their stories and anecdotes from their working lives were the stuff of my childhood. Several million Edwardian women worked in other people's houses for wages – a shared experience of life that I found surprisingly moving.

Looking to the future, I already have a number of Edwardian projects that I will be carrying forward. Some are practical, such as knitting a true Cornish Gansey in the round with a knitting sheath. When I first saw the boy's Ganseys I was deeply impressed, and having learned a little about them I was keen to have a go. It took me a while to get the knack of the knitting sheath but now I am click clacking away. At the speed I knit it will be a while yet before anyone will be wearing it. I have started looking into the early cosmetics business; I would like to spend more time on this and especially the 'culture of the skin'. I have no doubt that I shall also be returning to the whole subject of privies.

I can't end without mentioning those people who were behind the camera throughout our Edwardian year, such a thoroughly great bunch of people. It has been an added pleasure to be able to share the year with such hardworking, warm and funny men and women. Thank you, all of you.

ALEX

If there is one thing that my year spent on an Edwardian farm has brought so resoundingly to my attention, it is the issue of 'skills'. Throughout the year we have come across a vast array of highly skilled people and it has been a pleasure to learn from their dedication and admire their brilliance. For our project the question is most pertinent as it was in the Edwardian period that an ever diminishing pool of rural skills was being most keenly felt. Much was being made of the 'flight from the country-side' in the writings of the day and books such as *The Tyranny of the Countryside* (F. E. Green 1913), *Where Men Decay* (D. C. Pedder 1908) and *Folk of the Furrow* (C. Holdenby 1913) both lament the phenomenon and seek to explore the root causes behind it. Undoubtedly, there are a whole host of reasons why so many people left the countryside during the late-Victorian and Edwardian period but key among these were the lack of opportunities, the squalid housing conditions and a total lack of recognition, in financial terms, of the skills required to get the most out of a working countryside. Put simply, people who worked in the countryside were generally considered as 'unskilled' – that is, mere labourers – and were, as such, paid some of the lowest wages in Britain.

There is a huge irony in this for me because, in a day and age when we are going to be asked increasingly to turn to the local countryside to provide our food and basic sustenance, the skills so poorly rewarded over one hundred years ago will be the very skills we will need to achieve this shift. Expertise in growing food, tending livestock and managing the natural resources around us will fast become highly regarded, if the continuing pressure placed on global resources forces us to turn to the surrounding fields, pastures and meadows for the provision of our food and other material require-ments. It's an old cliché that the 'old-ways' are lost or dying but, as the boundaries of what is considered 'traditional' shift from generation to generation, we realize that Britain's countryside is that glorious mix of both the 'ever-changing' and the timeless. For us, our core work for the year had been the farm and engaging with a raft of skills that were demonstrably centuries, if not millennia old. But as we explored the other industries of the Tamar Valley, the remnants of the once vibrant mining landscape and the ghost of the terraced market gardens swiftly reminded us that the countryside is a place of change and one constantly in demand of new skills and fresh vision.

PETER

Blood, sweat, tears, broken bones, frustration, joy and a plethora of predominantly pleasant memories – this is the result of investing a year of

Sheepshearing was just one of the great experiences for Peter and Alex on the farm.

one's life, living away from friends and family, partaking in an historical farming project. I count myself very fortunate to have the opportunity to do this. Each time I watch the programmes I always think about the effort that goes in to every one of them not just from Alex, Ruth and I but from our wonderful crew, all our animals and the people that we meet along the way.

Our time in the West Country was a tumultuous one testing our mettle to the point of breaking. But, we survived and live to tell the tale of the wonders of Dartmoor and the farming particular to that untamed landscape. To recount adventures underground in the mines that lie like catacombs beneath our feet. To bear witness to what it was like to make a living from the sea and the foreshore, and as always experience what it was to farm in this area.

However, we are lucky in that we are only here for a year and ultimately our livelihoods do not depend on the outcome of our endeavours. Therefore, we can but glimpse into the true nature of the hardships, the daily grind and the camaraderie that was rural Edwardian Britain. Furthermore, if I were writing this round up of my farming year in 1914 I may well be penning this piece from a freshly dug trench in France hoping that I would be back before Christmas. The hardships that would follow in those war years would certainly make the previous fifteen years seem like a joy.

Personally, I feel that our time spent down in Devon with the occasional (if not frequent) foray into Kernow has been immensely successful. My hatchery, the lime kiln and charcoal burn, exploring the mine, roaming Dartmoor; these memories will stay with me till the day I die. So it is with a heavy heart and a tear in my eye that I leave this land and say goodbye to all who have helped us on our journey; to the cows, the sheep, Cyril our ram, King David our bull, our shires Prince and Tom, Jack the Clydesdale, Twinkle, Pickles and Laddie the ponies, the geese, the chickens, the pigs, the goats, the fish, the people. You shall be missed.

Acknowledgements

Alex and Peter would like to thank, in no particular order, the following list of people for their invaluable support and expertise throughout the year:

The production team and crew from Lion Television
Tim Hodge
Megan Elliott
Sara, David and Harry Birt
Francis and Janet Mudge
Colin and Hazel Pearse
Simon Summers
Rick 'The Pick' and Phil 'The Drill'
Chris Groves
Anne and Will Williams
Charlotte Faulkner
Jim Hamilton
Joshua Preston
Anthony Powers
Colin and Dan Richards
Mick Krupa
Lucy, Marion and Heather Langsford

Ruth would like to say an enormous thank you to all those people who so kindly taught her something of the West Country life.

Lastly – but far from least – David Upshal would like to thank the production team behind the television series whose hard work and tireless commitment throughout a long year made everything possible – Naomi Benson, Claire Smith, Felicia Rubin, Giulia Clark, Chris Mitchell, Tom Pilbeam – and, in particular, the genius that is Stuart Elliott who did such a magnificent job of leading and inspiring the team. Plus Matthew Nicholson, Lauren Jacobs, Liana Stewart, Faaiza Ahmed and Imogen Walford for being part of the gang at various stages of production. A huge thanks also to Laura Rawlinson for her exceptional photographs. And to our book agent Julian Alexander for his wise guidance.

BELOW Laura Rawlinson – in action – scrambling to capture the moment.
OVERLEAF Ruth, Alex and Peter with the amazing Bleriot aeroplane.

Further Reading

Arthur, Max. (ed.) *Lost Voices of the Edwardians* (2006)

Beckett, J. & Cherry, D. (eds.) *The Edwardian Era* (1987)

Bentley, N. *Edwardian Album: A Photographic Excursion* (1974)

Booker, F. *Industrial Archaeology of the Tamar Valley* (1967)

The Best Way, compiled from *Woman's World*, 1907

Conan Doyle, A. *The Hound of the Baskervilles* (1902)

Crossing, W. *Guide to Dartmoor* (1909)

Dale. R. & Gray, J. *Edwardian Inventions 1901–05* (1979)

Gardiner, J. *The Edwardian Country House* (2002)

Green, F. E. *The Tyranny of the Countryside* (1911)

Green, Jeffrey. *Black Edwardians: Black People in Britain, 1901–14* (1998)

Hasbach, W. *A History of the English Agricultural Labourer* (1894)

Holdenby, C. *Folk of the Furrow* (1913)

Horn, P. *The Changing Countryside in Victorian and Edwardian England and Wales* (1984)

Horseman, G. *Growing Up Between 1900 and 1920* (1996)

Hoskins, W. G. *Devon* (1954)

Howkins, Alun. *The Death of Rural England: A Social History of the Countryside Since 1900* (2007)

Langlands, A., Ginn, P. & Goodman, R. *Victorian Farm: Rediscovering Forgotten Skills* (2008)

Paige, R. T. *The Tamar Valley and It's People: The Years of Change 1840–1940* (1984)

Pearse, C. *Mill to Mill and Stook to Flour* (2010)

Pearse, C. *The White-Faced Drift of Dartmoor's 'prapper' Sheep* (2004)

Pedder, D. C. *Where Men Decay* (1908)

Read, D. *Documents from Edwardian England 1901–15* (1973)

Read, D. *Edwardian England: 1901–15 Society and Politics* (1972)

Reynolds, Stephen. *A Poor Man's House* (1909)

Rider Haggard, H. *Rural England* (1902)

Swan, E. (ed.) *William Carter Swan's Diary of a Farm Apprentice 1909–1910* (1984)

Winter, G. *A Country Camera 1844–1914* (1966)

Picture credits

All photographs taken by Laura Rawlinson, except for the following:

Page 4 (bottom): Tom Pilbeam; **Page 12:** Stuart Elliott; **Page 23:** Alex Langlands; **Page 25:** 'The Source of Unemployment' from The Museum of English Rural Life; **Page 37 (top):** Tom Pilbeam, **(bottom)** Giulia Clark; **Page 48–9:** Matthew Nicholson; **Page 52:** Alex Langlands; **Page 60:** Matthew Nicholson; **Page 61:** Alex Langlands; **Page 86 (bottom):** Alex Langlands; **Page 90 (bottom):** Tom Pilbeam; **Page 97:** Matthew Nicholson; **Page 99:** Matthew Nicholson; **Page 100:** Matthew Nicholson; **Page 101:** Stuart Elliott; **Pages 102–3:** Giulia Clark; **Page 107:** Giulia Clark; **Pages 110–14:** Matthew Nicholson; **Page 127:** © Mary Evans Picture Library; **Page 135:** Alex Langlands; **Page 138:** © Mary Evans Picture Library; **Page 153 (top right):** © Mary Evans Picture Library; **Page 162:** Stuart Elliott; **Page 167:** Matthew Nicholson; **Page 185:** © Mary Evans Picture Library/The Women's Library; **Page 187 (top):** Stuart Elliott, **(bottom)** Matthew Nicholson; **Pages 188–9:** Stuart Elliott; **Page 193:** Stuart Elliott; **Pages 198–9:** Matthew Nicholson; **Pages 212–13:** Stuart Elliott, **Page 214:** © Mary Evans Picture Library; **Page 215:** Lauren Jacobs; **Page 218:** Tom Pilbeam; **Page 229:** Tom Pilbeam; **Page 231:** Giulia Clark; **Page 232:** Giulia Clark; **Page 241 (top):** © Mary Evans Picture Library, **(bottom)** Giulia Clark; **Page 243:** Stuart Elliott; **Page 246:** © Nick Todd/Alam; **Page 247:** Stuart Elliott; **Page 249:** Giulia Clark; **Pages 263–5:** Tom Pilbeam; **Pages 267–8:** Matthew Nicholson.

Courtesy of the Lion Television crew:
Page 7, 13, 18–19, 38–39, 40, 41, 43, 45, 50, 53, 54–55, 56, 63 (both), 73, 74, 77, 80 (both), 82–83, 84, 85, 95, 120, 133, 141, 142, 143, 144, 146–147, 151, 156, 159, 162 (right), 166, 175, 176, 186, 191, 201, 202, 204, 210, 217, 222, 234, 235, 236, 251, 260–261

BELOW The Lion Television crew, filming rain or shine, day and night.
OVERLEAF The crew enjoying the end of year celebrations.

Index